AFTER THE FUNERAL

AFTER THE FUNERAL

The Posthumous Adventures of Famous Corpses

Edwin Murphy

A Citadel Press Book
Published by Carol Publishing Group

A Citadel Press Book
Published by Carol Publishing Group
Citadel Press is a registered trademark of Carol Communications, Inc.
Editorial Offices: 600 Madison Avenue, New York, N.Y. 10022
Sales and Distribution Offices: 120 Enterprise Avenue, Secaucus, N.J. 07094
In Canada: Canadian Manda Group, One Atlantic Avenue, Suite 105, Toronto, Ontario M6K 3E7
Queries regarding rights and permissions should be addressed to Carol Publishing Group, 600 Madison Avenue, New York, N.Y. 10022

Carol Publishing Group books are available at special discounts for bulk purchases, sales promotion, fund-raising, or educational purposes. Special editions can be created to specifications. For details, contact: Special Sales Department, Carol Publishing Group, 120 Enterprise Avenue, Secaucus, N.J. 07094

Manufactured in the United States of America
10 9 8 7 6 5 4 3 2 1

Library of Congress Cataloging-in-Publication Data
Murphy, Edwin.

   After the funeral : the posthumous adventures of famous corpses / by Edwin Murphy.
       p.   cm.
   "A Citadel Press book."
   ISBN 0-8065-1599-6 (pbk.)
   1. Celebrities—Death—Miscellanea. 2. Celebrities—Biography—Miscellanea. 3. Dead—Miscellanea. I. Title.
   CT105.M87   1995
   920.02—dc20   94-45464
   [B]   CIP

To Richard Roth

CONTENTS

PREFACE

Most biographies stop with the death of their subject. After a death scene, and occasionally a brief mention of the funeral, biographers usually think they have finished their task of portraying whomever they are writing about. If anything occurred after the funeral, they are unaware, don't care, or think it irrelevant. However, many amazing things happen to the dead bodies of famous people, and biographers do a disservice to their readers by failing to include these incidents, which often are directly relevant to fully understanding a person's significance in history.

Ending a biography at the graveyard, with the tacit implication that the subject will then rest in peace, can be misleading. "Rest in peace" is an evasive and unsatisfying way to end a biography, no more informative than ending a fairy tale with "they lived happily ever after." Just as in real life people don't always live happily ever after, so too in real death, famous people's bodies are not always allowed to rest in peace. In fact, research shows that the more notable a person was during life, the less likely it is that he or she will be allowed to rest in peace. The living are inclined to disturb the remains of their celebrated predecessors for all kinds of reasons. Love, hate, respect, revenge, worship, curiosity, sentimentality, politics, science, humor, greed, and many other motives have been at work, not to mention pure accident, in keeping the distinguished dead from enjoying their repose.

After the Funeral attempts to remedy the oversight of many biographies. It continues where they leave off, telling the shocking, surprising, touching, humorous, and other interesting things that have happened to some eminent corpses throughout history. In these pages, you will read of the Portuguese concubine who was exhumed five years after death to be crowned queen of her country. You will discover that Oliver Cromwell, Lord Protector of England, was dug up years after he died so that he could be hanged for treason. You will read about the executed labor activist whose cremated ashes accidentally found their way into the United States Archives. You will be amazed to learn that Albert Einstein's brain was lost for over twenty years after the autopsy and that Voltaire's heart was misplaced for decades in the Bibliothèque Nationale in Paris.

There are so many outrageous stories of this sort in *After the Funeral* that I fear readers will find some of them hard to believe. At least, they may suspect that I am exaggerating or repeating unsubstantiated rumors, especially since most of the episodes I included are not well known. For instance, one may consult twenty standard biographies of Cardinal Richelieu, as I did, without finding the least reference to the surprising indignities that were inflicted on his detached face. Most biographies of Albert Einstein make scant reference, if any, to his misplaced brain. It is nearly always the same. Biographers are reluctant to write anything about their subjects' dead bodies, particularly anything that is embarrassing, ridiculous, or unseemly.

Therefore, I was forced to delve deeply and widely in "many a quaint and curious volume of forgotten lore" to find the information I needed for this collection of posthumous histories. Frequently, as in the case of D. H. Lawrence's lost ashes or Thomas Paine's stolen corpse, I had to piece together a coherent narrative from bits and snatches of relevant information scattered about in disparate sources. On many occasions I was faced with contradictory statements, such as the four versions of what happened to John Barrymore's corpse. When I could not reconcile discrepancies or separate the true from the false, I either cited all the variants or presented the most plausible one. For the benefit of the dubious, as well as of other researchers, I have included a complete bibliography of all the sources in which I found useful information.

The first thing people usually ask me when they hear about *After the Funeral* is "Where did you get the idea to write a book about dead bodies?" I am led thereby to believe that the answer might be of interest to readers.

On an evening in 1984, when my friend Richard Roth and I were working overtime and had stopped for a break, I began trying to interest him in writing a book. I had tried many times before, without success. Rick was one of the most erudite people I had ever known, with encyclopedic knowledge of biography, history, royal families of the world, art and art history, music, philosophy, religion, the occult, and many other subjects. Yet he tended to keep his wisdom to himself, opening up only to a few close friends. His learning was so extensive, his understanding so deep, his perceptiveness so acute, that I thought he should share some small fraction of his accumulated knowledge with the world, in the form of a book. That night he finally agreed, but on the condition that I collaborate with him on the project.

We began casting about for an interesting and original subject. Since both of us knew numerous anecdotes about unusual things that had happened to the dead bodies of famous people, and since this was a topic not previously attempted by other writers, we decided to make our book an anthology of such stories, sort of an exercise in necrobiography. Such was the genesis of the volume that you hold in your hands today and that you are eager to start reading if only I would finish this preface.

Alas, work on our book was delayed several years by a literary project to which I was already committed. Then, before Rick and I could get beyond the most preliminary research on *After the Funeral*, he died suddenly. I determined to write the book myself and dedicate it to Rick, my friend. Ultimately, this took longer than expected, but here it is at last. I owe Rick the credit for helping find such an interesting and original theme, for making valuable suggestions about the organization and structure of the material, and for regaling me with several of the choicest anecdotes covered in the following pages. I hope I wrote them with as much wit and skill as he told them.

When I started this book, I imagined that I would be able to find a limited number of interesting stories about the dead bodies of famous people. Instead, I found hundreds, not all of which I was able to include in one volume. I hope my readers will find *After the Funeral* so interesting that they will clamor for a sequel. Anyone who wishes to tell me about a good "dead body" story I may have overlooked, or who has additional information about the people included in this volume, may contact me through the publisher.

ACKNOWLEDGMENTS

My wife, Valérie, gave me untiring assistance in many phases of this work, including correspondence, research, French and German translating, typing, file organization, and valuable editorial and content suggestions. She gave me all the support and encouragement I could desire. I owe thanks to my twin brother, Thomas Murphy, for giving me some ideas and for translating from Spanish Eusebio Leal Spengler's lengthy research paper on Columbus; to Richard C. Baker, who expertly translated, from medieval Portuguese, the relevant section of *Cronica de el-Rei D. Pedro I* by Fernão Lopez; and to Jeff Hearn, who indefatigably tracked down many useful sources of information on the bodies of Abraham Lincoln, Sitting Bull, Eva Perón, and others. Finally, in addition to my indebtedness to Richard Roth, I am grateful to my friends Michael Reinhardt, Tom May, John Johnson, and Jim Bond for giving me leads on some "dead body" stories I might not otherwise have found.

PART I

HEADS

A Word About Heads

The head, more than any other part of the living body, is what we associate with the whole person: the physical identity, so to speak, of the individual, rolled up into a compact little sphere. It contains the face, a human's most distinguishing feature, the one most often observed and most likely to be remembered. When we think about a fellow mortal, nine times out of ten it is the face that distinguishes one particular person, in our mind's eye, from all others. In our mental images of other people, as in portraits, the torsos, hands, feet, arms, and legs are essentially interchangeable, except in unusual cases. But the face is a unique marker denominating the physical distinctness of one human being from every other.

Also, the head contains the brain, which is the center of a person's thoughts, emotions, behavior, and personality. Some cultures even have equated the mind with the soul, and consequently they regard the brain as the locus of the spiritual existence of each individual. The head, in this concept, is the home of the soul, at least symbolically.

For the foregoing reasons, dead heads often are treated differently from dead bodies. Heads have the additional practical advantages of being light and compact, easily detachable, easy to store and transport, preservable (especially the skull), identifiable, and, at least to some people's tastes, decorative. These advantages are especially pertinent to the practice of head-hunting. The deliberate acquisition of someone else's head, usually that of a rival or an enemy, for magical, ritual, or occult purposes is a venerable custom once found in diverse cultures around the globe. Today it is practiced routinely only in the remotest corners of Borneo, New Guinea, and the Amazon basin, but it used to be a proud tradition among the warrior classes of many nations.

In the most obvious sense, bringing home the head of an enemy demonstrated conclusively who was the stronger in combat. However, this was seldom the only reason, or even the primary reason, for taking someone's head. Many primitive societies placed some degree of magical significance on forcibly acquiring the head of an enemy. By possessing his head, you possessed or controlled his spirit; you possessed his power, his luck, his merit, and his prestige, and you added these valuable attributes of his to your own accumulated store of them. The heads stuck up on poles around your house, shrunken and hung on your belt, or kept on a shelf in your banqueting hall were much more than ostentatious trophies of your martial prowess. They were evidence of your acquired psychic, moral, and magical power and of the augmentation of your spiritual capital at the expense of others.

For the last hundred years or so, commerce has helped support the art of head-hunting in certain remote areas where it was in decline. Tourists, anthropologists, and museums seeking shrunken heads for their collections, and offering handsome prices for them, create a demand which indigenous peoples, conforming to the economic dictates of modern civilization, are willing to supply. For example, when Michael Rockefeller, great-grandson of billionaire John D. Rockefeller, arrived in the wilds of New Guinea in 1961, collecting artifacts for the family museums, the Dutch colonial administrators were not amused. They had finally convinced the frontier tribes to give up head-hunting (or so they thought), and now this rich playboy was offering ten steel knives apiece for suitable specimens of the headhunters' art. Knowing it was illegal to snatch heads, but tempted by the obvious value the white stranger placed on them, some morally perplexed warriors applied to the authorities for special permission to go head-hunting "just for one night." Permission was denied. Rockefeller disappeared after a boating accident a few days later and was never seen again.

As societies progressed beyond their primitive stages and the old belief in magic declined, head collecting for occult purposes ceased. Heads continued to be sought, especially in war, but now for more pragmatic reasons, such as to inspire terror in the enemy. This was the fate of Blackbeard the pirate, whose head was cut off after his defeat in battle with the authorities, then paraded through the ports of North Carolina and Virginia and finally stuck up on a pole near the Hampton River as a warning to other pirates. It later reputedly served as the base of a punch bowl at Raleigh Tavern in Williamsburg, Virginia. Sometimes a head was used as conclusive proof that an enemy was really dead, as when, in 1885, General Charles Gordon's head was delivered to his enemy, the Mahdi, after the fall of Khartoum in the Sudan. Or the object may simply be to add insult to injury, such as in the case of Rafael Trujillo, strongman of the Dominican Republic for thirty years: he was assassinated by a band of gunmen, who absconded with his head (the rest of him is buried in Paris).

The ancient Celts delighted in making cups, lamps, and other grisly utensils out of human skulls, as did many nomadic tribes of central Asia. Even Napoleon was said to own a goblet fashioned from the skull of Cagliostro, the notorious Italian adventurer. The Turks, as late as the eighteenth century, frequently erected towers built from the skulls of rebels as grim warnings to others of the dangers of revolt. Adahoozou I, king of Dahomey in the sixteenth century, lined the walls of his

5

capital, Colmina, with the heads of his subjects, some requisitioned especially for this awe-inspiring purpose.

In certain societies, such as the Celtic and the Aztec, the heads of sacrificial victims played a special role in religious ritual. The Celts even thought that nothing guaranteed the sweetness, purity, and unfailing abundance of a fresh water spring like ceremonially dropping a suitably sanctified human skull down to the bottom of it. Perhaps that is where the term *headwaters* originated.

Other societies, including the Christian, have preserved and venerated the heads of their distinguished men and women. European churches and monasteries frequently contained shrines and reliquaries displaying the heads of saints for the edification of the faithful. In our unfortunately more secular era, the skulls of famous people are still put on display, in museums rather than churches, and for the edification of the curious rather than the pious. The heads of Haydn, Mozart, Cromwell, and Phineas Gage, as we shall see, all served time in museums. Let's go on to discover what strange adventures befell these and other famous heads, *after the funeral*.

CHAPTER 2

Who Hid Haydn's Head?

The world-famous composer Joseph Haydn lost his head in 1809. To be exact, it was stolen. Fortunately, he was dead at the time. Haydn wasn't recapitated until 1954. In the intervening years, 145 to be exact, his head, whose brain enriched the world with imperishable music such as *The Creation* and *The Seasons*, was treated as a prize over which men, museums, and countries connived and fought like naughty playground children squabbling over a pretty marble. The deceased maestro's cranial unit was stolen, hidden, substituted, lent, bequeathed, displayed, and passed around so often that it makes your head spin to think about it. Such are the indignities that can be heaped upon the remains of even such a beloved, gentle genius as Joseph Haydn, *after the funeral.*

As usual in tawdry cases such as this, the complete truth is hard to come by. Besides the ordinary problems of sparse documentation, conflicting testimony, poor memory, self-interest, and the reluctance of biographers to mention disagreeable facts about their subject's dead body, there is the additional difficulty of the deliberate lies and deceptions which encumber this story filled with multiple thefts and unseemly disputes. The following narrative is pieced together from many disparate sources, none of which individually gives the entire picture, and all of which disagree to some extent on names, dates, sequence of events, and circumstances.

That being said, what the heck happened to Haydn's head? The renowned composer died at his home in Vienna on May 31, 1809, of old age (he was seventy-seven). It was an

Joseph Haydn's head was stolen by phrenologists. (*Courtesy of Library of Congress*)

inconvenient time to have a fancy funeral, since the Austrian capital had just been occupied by Napoleon's troops. So, despite Haydn's immense popularity, nothing much was done. On June 1, his corpse was quietly removed to the parish church of Gumpendorf in an oaken casket. After the funeral on the same day, it was buried in the Hundsthurm (or Hundsturmer) cemetery at the gates of Vienna.

It didn't stay buried long. The enemy occupation, it seems, had not dampened the eagerness of certain Viennese phrenologists to examine the bumps on famous skulls. A cabal of them bribed or persuaded Joseph Rosenbaum, secretary to Prince Nicolaus II Esterhazy, Haydn's patron, to steal the musical genius's head and lend it to them for a few weeks. Rosenbaum, a dabbler in phrenology himself, was undeterred by his close friendship with Haydn. He and an accomplice bribed the cemetery employees to look the other way while they secretly exhumed the body just days after interment, cut off the head, and reburied the rest.

The phrenologists, a group of doctors, eventually returned the head to Rosenbaum, who kept it. The theft might have gone undetected forever, except that in 1820 a visiting Englishman reminded Prince Esterhazy of something he had neglected far too long. He had intended to transfer Joseph Haydn's corpse to a specially built mausoleum on the family estate of Eisenstadt, where the great music master had lived and worked for the prince's grandfather for thirty years. He immediately ordered the body disinterred. To everyone's horror, it was a head short!

The prince was apoplectic with rage. The police were called in, and Rosenbaum was implicated. The police searched his house but failed to find the missing cranium, which Frau Rosenbaum hid under the mattress of her feigned sickbed. Sorely in need of money, Rosenbaum and his accomplice negotiated. If the prince offered a suitable "reward" for the missing head, perhaps it would turn up. Esterhazy correctly viewed the proposed reward more as a bribe or a ransom, which didn't bother him much since he never got around to paying it anyway; but he promised a tidy sum. The grave robbers, quite rightly, didn't trust the prince to keep his word, so they palmed off a substitute head on the gullible nobleman, who, not being a phrenologist, could not tell one skull from another. So both parties got exactly what they deserved from this infamous bargain.

Prince Esterhazy didn't learn until later that the head he reverently reburied in Haydn's new vault at Eisenstadt was a fake. Rosenbaum confessed on his deathbed, but still he wouldn't give the real skull to the prince. He formally willed it to the Vienna Gesellschaft der Musikfreunde (Society of the Friends of Music, also known as the Academy of Music). Unfortunately, it was no longer his to bestow. It had been stolen again.

This time the culprit was the doctor who attended Herr Rosenbaum in his last illness. He swiped the skull on the very day his patient expired, perhaps to make sure of his fee (dead patients are notorious deadbeats) or perhaps because he too was a dilettante phrenologist. In any case, he eventually sold the criminally acquired cranium to a noted Austrian professor, who died with it still in his possession. He bequeathed it, or, as some sources report, his widow gave it to the Pathological Museum of the University of Vienna.

When the Gesellschaft der Musikfreunde got wind of the Pathological Museum's prize acquisition, it sued to recover what it considered its rightful property: after all, the first thief had willed the skull to it, fair and square, before the second thief stole the contraband from the first thief and sold it to the professor who passed it on to the pathologists. Is that clear?

The court thought it was, but only after protracted and costly legal processes. The court also brushed aside the claim of the Esterhazys, who, finally realizing they had been duped, tried to enter the case on behalf of the deceased, whose body they possessed and who, nobody could deny, was the original owner of the head.

The Gesellschaft der Musikfreunde finally got possession, legally, in 1839 and eventually decided to put Haydn's skull on display in its museum, where it rested from 1895 to 1954 in a glass case sitting on a piano. Visitors were even allowed to handle the relic if they wished.

Still, the Esterhazy family never ceased to press its argument that Joseph Haydn deserved a better fate than to lie headless in his grave while his pilfered skull was shamelessly exposed as a museum curio for ogling tourists. Generation after generation, they persisted in their quiet attempts to buy, beg, or borrow poor Haydn's head for restoration to its torso in the tomb at Eisenstadt. The political situation didn't help. The faltering Austro-Hungarian Empire was finally dismembered after World War I, and Eisenstadt for a while was part of Hungary. The Nazi occupation and World War II were followed by Soviet occupation.

At last, in 1954, the Esterhazy family convinced the Gesellschaft der Musikfreunde to transfer the head to them for proper burial. As negotiations were nearing completion, however, the Communists in Hungary imprisoned the patriarch of the family, Prince Paul Esterhazy, who was a Hungarian citizen. They also confiscated the family property, including the estate at Eisenstadt in the Soviet-occupied part of Austria. Despite fear that the Communists might seize Haydn's skull if it were sent into the occupied province, the authorities of Austria decided to take the chance. The transfer would be a cultural, not a political, event, and therefore it was scheduled to take place during the Vienna Music Festival in June 1954.

On June 5, Whitsunday, the skull was blessed by the Cardinal Archbishop of Vienna at a ceremony attended by the president and the chancellor of Austria and foreign diplomats. Placed in an urn decorated with laurel wreaths and borne on a hearse, the revered head was carried in procession for thirty kilometers southeast through Burgenland to Eisenstadt, passing through Haydn's birthplace at Rohrau. Everywhere along the way, the streets were hung with flags and church bells rung.

Meanwhile, Joseph Haydn's headless trunk had been unearthed and placed in a new copper casket at the baroque-style Bergkirche (Hill Church) of Eisenstadt. When the head arrived, the prominent sculptor Professor Justinus Ambrosi placed it gently on its cushion in the casket, finally reuniting it with the body from which the grave-robbing phrenologists had severed it in 1809.

The complete corpse was reinterred in the Bergkirche after a fitting service. On the tombstone is written:

Non moriar—sed vivam et narrabo
opera Domini.

I shall not die, but live,
And declare the works of the Lord.

CHAPTER 3

Mozart's Head? Only His Gravedigger Knows For Sure

W olfgang Amadeus Mozart died on December 5, 1791. He was thirty-five years old. Many consider him the greatest musical genius of his day; others hold him the greatest of all time. His skull, like that of his friend Joseph Haydn, was destined to be a museum exhibit for many years; in fact, Mozart's cranium remains in a museum to this day. At least, *someone's* head bone resides in the museum of the International Mozarteum Foundation in Salzburg, Austria. Whether it once belonged to Mozart has long been a subject of debate. Several scientific teams claim the skull is genuine, while the Mozarteum maintains that there is no sufficient proof.

Let's see how this skull found its way to the Mozarteum and caused all this fuss. On the day after Mozart's death, his body was laid out and consecrated in a chapel, the Kreuzkapelle, of the great St. Stephen's Cathedral in Vienna. Only a few friends attended the rites, including his rival, Antonio Salieri, whom conspiracy theorists and Broadway producers sometimes accuse of having poisoned the young genius. Even Wolfgang's widow missed his funeral, being overcome with grief.

Wolfgang Amadeus Mozart's skull is in a museum in Salzburg—or is it? (*Courtesy of Library of Congress*)

In the Austrian capital, transporting dead bodies during daylight hours was prohibited by sumptuary and health ordinances, as were funeral processions and graveside ceremonies. Therefore, after the funeral service, Mozart's coffin sat waiting in the mortuary chapel, the Kruzifixkapelle, until nightfall.

Then, unaccompanied by mourners and followed only by a stray dog, a hearse trundled the body out beyond the eastern city wall to a cemetery near the village of St. Marx. It rested there with several others until the next morning, December 7, when two gravediggers removed them all from their reusable coffins and buried them in a common grave. Such communal interment was the normal procedure at the time; practically no one had the luxury of an individual grave, and gravestones were prohibited to save space and expense.

Although the gravesite was not indicated by a marker, one of the gravediggers, Joseph Rothmayer by name, supposedly remembered both the location of the plot and the placement of Mozart within the crowded grave. Ten years later, when that part of the cemetery was undergoing one of its periodic "reorganizations" (e.g., the removal of older bones to make room for fresh corpses), he rescued the decomposed composer's head and kept it as a memento, thus saving it from oblivion like the rest of the skeleton. Either Rothmayer or someone else scraped the remaining flesh and brain from the bone, as evidenced by striations on all surfaces of the skull. Was this actually Mozart's cranium, or someone else's? Only his gravedigger knows for sure.

Rothmayer left the skull to his successor in the job, Joseph Radschopf, who later, around 1842, gave it to an acquaintance named Jacob Hyrtl. When Jacob died in 1868, his brother, Joseph Hyrtl, inherited the skull. Joseph, by coincidence, was a prominent Viennese anatomist and phrenologist with a large collection of famous heads, including casts of the skulls of Beethoven and Schubert. As we saw in our discussion of Haydn's head, phrenology was all the rage in Vienna in those days. Probably for aesthetic reasons, Joseph removed the jawbone from the skull, and it has since disappeared.

Joseph Hyrtl's widow inherited his macabre collection when he died (one wonders what became of *his* skull); upon her own death, and the settlement of her estate on October 6,

1901, Mozart's skull was bequeathed to the Mozarteum in Salzburg, where it has been ever since.

Salzburg was Mozart's birthplace, and at last he had come home; or at least part of him had. Of course, the nagging question remained: was this really Mozart's skull? What if the gravedigger was mistaken, or just told a fib to impress his friends? Maybe this skull, still bearing Joseph Hyrtl's neat label on its forehead, actually is that of someone else, the village idiot perhaps.

Although lacking reliable proof that the skull belonged to Mozart, the Mozarteum directors began displaying it as his in 1902. They removed it from public view in the 1950s, partly because no conclusive evidence existed that it was genuine and partly because such an exhibition, acceptable in 1902, by mid-century was considered of questionable taste. Also, the directors of the institution were getting wary of the unpredictable effects the skull produced in many of their visitors. Some were overcome with emotion and wept at the sight, while others scoffed and called it a hoax.

Several attempts to authenticate the skull were made early in the twentieth century, but results were inconclusive. In the 1980s, a team of French anthropologists led by Pierre-François Puech of a university in Provence, using forensic analysis, claimed nearly positive identification of the skull based on portraits, descriptions, and the medical history of Mozart. The team employed "forensic reconstruction," using soft clay to model facial features of the deceased around the bones, and came up with what they considered a strikingly accurate likeness of Mozart. Unfortunately, the same technique, used a few years later by six scientists at the Vienna Museum of Natural History, resulted in another reconstructed head which, although its fabricators claimed it was the spitting image of Mozart, didn't look at all like the first.

The governing board of the Mozarteum had commissioned the latter study, in connection with the two hundredth anniversary of Mozart's death. Intending to refute the claim of the French team, the board instead found itself in the awkward position of having to disavow the results of its own investigation. According to the ultra-cautious members of the foundation, neither the French nor the Viennese study results are solid enough to prove conclusively that the skull is Mozart's, only that it *might* be. The Mozarteum board hurriedly called in a new consortium of seven Swiss and German anthropologists, neurologists, and forensic experts, who reviewed the work of their predecessors and obligingly pontificated that some of their techniques, especially facial reconstructions, were unreliable as hard evidence.

As new scientific techniques come along, the Mozarteum will continue trying to get ironclad proof that their prize skull is, or isn't, that of the musical genius. Meanwhile, they keep lending the cranium to medical researchers whose morbid hobby is trying to establish the real cause of death of famous figures like Mozart. In this case, as in most such cases, there are as many theories as there are researchers. This game has two rules: (1) never accept the testimony of physicians who treated the deceased or performed the autopsy, because they are always wrong about the cause of death; (2) never accept the conclusions of all your predecessors in the game, but come up with some previously undetected cause of death completely overlooked by everyone else. Following these two rules almost always leads to a book contract, a lot of journalistic publicity, and brief fame for the forensic detectives until someone else comes along in turn with a new theory to debunk theirs.

Mozart's doctors described his last illness as *hitziges Frieselfieber*, "high fever with rash." Other theories through the years variously held that Mozart was killed by tuberculosis, rheumatoid arthritis, streptococcal infection, or poisoning by Salieri or by the Freemasons. The research of a French physician, J. Barraud, in 1905, led to the standard modern diagnosis that Mozart died of uremic poisoning caused by progressive kidney failure. The International Mozarteum Foundation officially backs the theory of Carl Baer, published at its expense in 1972, that the real cause of death was congestive heart failure resulting from the chronic effects of rheumatic fever. The French team led by Puech, based on analysis of the skull, contends that death was occasioned by chronic subdural hematoma, or leakage of blood into the space between the brain and skull. A prominent crack in the skull's left temple, only partly healed at the time of death, supports this hypothesis if, of course, the skull really is Mozart's.

Miles Drake of Ohio State University Hospital analyzed the skull in 1993 and agreed that subdural hematoma was present, but that it was not directly responsible for the composer's untimely demise. That result was accomplished by his doctors, who bled him so profusely to combat the attendant fever that a sudden drop in blood pressure probably caused a stroke.

Whatever the truth, the skull reputed to be that of Wolfgang Amadeus Mozart remains safely in the custody of the Mozarteum, awaiting the next set of researchers, and the next, and the next. At least they all agree that Mozart is dead. But his divine music lives on, unaffected by the fate of the mute cranium in Salzburg from which it may have emanated. Wolfgang Amadeus, "beloved of God," has gone to meet his Lord, and left his heavenly music for the ages of man.

The Crazy Career of Cromwell's Cranium

O liver Cromwell, Lord Protector of England, scourge of the Cavaliers, the Catholics, the Irish, and just about anyone else who disagreed with his religion or his politics, had the good fortune to die peacefully on September 3, 1658, shortly before the restoration of the monarchy. Judging from what the royalists inflicted on his corpse, Cromwell's death might have been a lot less peaceful if his enemies had gotten their hands on him while still alive. But they didn't, and he was quietly embalmed and privately buried in Westminster Abbey, with his head still on his shoulders. A lifelike wax effigy was laid in state and given an elaborate public funeral several months later, to amuse the masses.

The monarchists, however, were not amused to be cheated of their revenge for the regicide of Charles I in 1649. Soon after regaining power in 1660, they exhumed the real Cromwellian cadaver, along with those of two associates, for a ludicrous public vengeance. On January 30, 1661, after spending a night under guard at the Red Lion Inn at Holborn, Oliver Cromwell's body was dragged openly through the London streets to Tyburn, the place of punishment for common criminals. At ten o'clock, still wrapped in its grave clothes, it was given a symbolic

19

execution by hanging. Six hours later, the executioners lowered the corpse from the gallows and beheaded it, a bungling operation that required six or eight blows of the ax. A few souvenirs were snipped off at this time for personal keepsakes, such as fingers, toes, and an ear, but nothing more is known about what happened to these crumbs of Cromwell.

The major fragments, head and body, underwent the usual treatment meted out to traitors. The "odious carcase," as a contemporary record calls it, was buried in a deep, unmarked pit beneath the scaffold. The head was paraded through the streets for the amusement of the crowd. The same populace which two years earlier wept and gawked at the funeral of Cromwell the Lord Protector now hurled insults and garbage at the head of Cromwell the traitor. Five days later, the head appeared prominently displayed atop Westminster Hall, affixed to an iron-tipped pole driven through the crown of the skull. There it remained for many years, as a caution to would-be rebels.

The wind blew the head down one night in 1685. A sentry secreted it beneath his cloak and took it home. Then began its long wandering, passing from hand to hand and indignity to indignity before obtaining another decent burial almost two centuries later.

The complete history of the head's peregrinations is not known. There are gaps in the record, but the head kept popping up at irregular times and places, and its authenticity is fairly certain. Forensic studies reinforce the obvious fact that the embalmed head, even after considerable wear and tear in its many adventures, still looks like Oliver Cromwell, right down to the prominent wart over one eye that is so visible in all of his portraits. The age, size, shape, facial features, hair, missing ear, and trepanned cranium (a feature of the autopsy, recorded in detail) all match the known facts. Even the imputed brain mass of 1,500 cubic centimeters generally fits the 2,126 grams noted in the autopsy, and the six ax wounds on the neck and the flattened nose correspond exactly to descriptions of the difficult decapitation.

No two sources give exactly the same sequence of events concerning Cromwell's head after the sentry purloined it as a touching memento of his dreary employment. The following narrative is a collation of several detailed descriptions, all of which agree in broad outline, but which vary considerably in the details included, the dates, and the order of events.

It seems the sentinel's daughter at some point sold the souvenir to a family in Cambridgeshire. Later, from 1710 to 1738, it was located in a private London museum. At an uncertain date it was acquired by the Russell family, and eventually, between 1773 and 1780, descended to the possession of Samuel Russell, a poor, second-rate actor at Covent Garden. He also operated a small museum near Long Acre, London, in which Oliver Cromwell's upper story became the prime exhibit.

Russell once tried to sell the trophy to Sidney Sussex College, Cambridge, where Oliver had been a scholar during one year of his young life. The college authorities declined the offer. They still held a grudge against Cromwell for melting down their silver plate to finance his war against Charles I and for imprisoning their headmaster. Another private museum impresario, the jeweler James Cox, at last bought the head from Russell, and soon he resold it to a trio of speculators for £230, a handsome price in those days. By 1799, it was again on public display, this time in a Bond Street exhibit.

In 1814, Josiah Henry Wilkinson bought the restless head. He used to take it to parties as a conversation piece and titillate the other guests by passing it around, iron spike and all, while giving lectures on phrenology. One disgusted partygoer in 1822, Maria Edgeworth, confided in a letter:

Mr. Wilkinson its present possessor doats on it—a frightful skull it is—covered with its parched yellow skin like any other mummy and with its chestnut hair, eyebrows and beard in glorious preservation—The head is still fastened to a pole. Mr. and Mrs. Ricardo and the family by turns held the head . . . while the happy possessor lectured on it compasses in hand . . . To complete Mr. Wilkinson's felicity, there is the mark of a famous wart of Oliver's.

Cromwell's head stayed in the Wilkinson family for almost a century and a half. Josiah's heirs and descendants began to treat it with more respect. It was subjected to various antiquarian and scientific investigations from time to time, being exhibited before the Royal Archaeological Institute in 1911. Two "cranial detectives" studied it thoroughly in the 1930s and concluded that it was authentic beyond all reasonable doubt, a finding endorsed in later analyses of the evidence by the surgeon Dickson Wright and others.

At length the head descended to Canon Horace Wilkinson. Upon his death, he bequeathed it to Sidney Sussex College, whose earlier decision not to purchase this relic of its errant scholar from Samuel Russell was abundantly justified by its now acquiring the same item as a gift.

In 1960, the college decided to give Cromwell's battered head a decent burial. There being no way to find, much less identify, Cromwell's body beneath the busy streets of London, the head was interred somewhere in or near the entrance to the college chapel. The exact spot is a well-guarded secret, to prevent any future desecrations at the hands of pranksters, collectors, or Irish extremists still angry over Cromwell's treatment of their ancestors at Drogheda and Wexford in 1649.

Cardinal Richelieu: Losing Face

A contemporary of Oliver Cromwell, across the Channel, was the great statesman Cardinal Richelieu (Armand Jean du Plessis) (see plate 3), who, as Louis XIII's minister, ruled France with an iron will for eighteen years. His diplomacy shook the very foundations of Europe, and at his death on December 4, 1642, he was at the height of his power. Like Cromwell, however, his power during life did not protect his body from desecration after death, nor prevent his head from being passed around like a football on a Sunday afternoon.

As part of the embalming procedure, the front half of Richelieu's head was sawed off for removal of the brain and other matter. The front and back portions of the skull were then clamped back together for the solemn funeral. After four days of lying in state at the Palais-Cardinal, his body was transferred with great pomp to the chapel of the Sorbonne, which he had built, and was interred with somber ceremony. Near the end of the century, a beautiful funerary statue was erected over the cardinal's grave.

Richelieu was undisturbed in his tomb until after the French Revolution. The revolutionaries, however, considered him one of the greatest villains in the history of France, unfit to be remembered. Thus, in December 1793, workmen were engaged in exhuming his corpse, and those of other undesirable relics of the ancien régime, from their honored graves in the Sorbonne chapel. While the workers were at lunch, a citizen shopkeeper, a corset maker named Cheval, wandered into the church as part of a rowdy mob. Cheval, finding Richelieu's

Cardinal Richelieu's mummified face was a family heirloom—in someone else's family. (*Courtesy of Library of Congress*)

coffin unattended, broke off his well-preserved head, held it up for the amusement of his friends, then sneaked away with the front half of it under his coat. Later, the laborers buried the rest of the faceless cardinal's body in a hole in the church basement.

Cheval kept his prize at home for a while, but began to worry about the possible consequences of his act of desecration, especially if the tides of politics should chance to turn against the Revolution. He tried to get rid of Richelieu's head in various ways, perhaps trying to sell it, but the precise circumstances are poorly documented and the sources open to different interpretations. He is said to have given it to a certain "Abbé Boschamp," founder of the municipal library of Saint-Brieuc in Brittany, while that gentleman bibliophile was on a book-buying expedition to Paris. Boschamp, whom historians have had trouble identifying with certainty, allegedly gave the unseemly relic to Nicolas Armez. Some researchers think that Armez got the face of Richelieu directly from Cheval, of whom he was a client, and invented the intermediary role of the mysterious Abbé Boschamp to distance himself from Cheval's act of desecration. In any event, Armez was in possession of the object by 1796, at the latest.

Armez was an abbé from Brittany. The face of Richelieu remained in his family for over seventy years. After the restoration of the monarchy, if not before, the family held the curiosity in great honor, keeping it in a glass globe at their home, the Château du Bourblanc in Plourivo, Brittany. They used to display it, as a special favor, at the local *collège* on the days when academic honors were bestowed. In 1812 or 1813, Nicolas Armez noticed that insects had started to infest the face, so he had it treated by a pharmacist in Rennes to prevent further deterioration. About 1820, Armez and the descendants of Cardinal Richelieu began a series of contacts and negotiations for the restitution of the cardinal's severed face to his own family, but these came to nothing. Also about 1820, Nicolas gave the face to his brother, who ten years later lent it to a prominent phrenologist, Dr. Broussais, for study.

Louis-Philippe Armez, nephew of Nicolas, mayor of Plourivo and a deputy in the Assembly, eventually inherited the family treasure. In 1840, he lent it once again, this time to the artist Bonhomme of Paris, who had been commissioned to do a painting of Cardinal Richelieu for the Palais du Conseil d'État. Bonhomme used it as a model, for even after two centuries it still was in a remarkable state of preservation. Photographs taken about 1860 verify this, as do eyewitness descriptions such as this by Alexandre Lenoir, who had witnessed the desecration (quoted in *Causeries d'un curieux* by Feuillet de Conches, 1862):

> The Cardinal, whom I saw taken out of his coffin, had the appearance of a whole dry and well-preserved mummy. Dissolution had not changed his features. His skin had a livid color. He had sharp cheekbones, thin lips, and red body hair, and his hair was whitened by age. One of the partisans of the government of 1793, thinking he was avenging in his furor the victims of this cruel minister, cut off Richelieu's head and showed it to the bystanders who were in the church at that time.

Pressure continued to mount on Louis-Philippe Armez to relinquish his family heirloom to either the state, a suitable institution, or the Richelieu family, but he was reluctant to part with it. When the Revolution of 1848 broke over the nation, he transferred the visage of Cardinal Richelieu for safekeeping to an unspecified but "trustworthy" friend, probably Auguste de Keratry, deputy from Finistère. Armez declared, "My intention is not to relinquish it . . . particularly not to see it buried inside a tomb," because that would be "as if this head did not exist." Armez temporized for twenty-two more years, despite increasing demands by the government. In 1854, he even politely refused a hint of the Emperor Napoleon III, who was instigated to intervene in the affair by the followers of the Comte de Montalembert, a leader of the liberal Catholic movement.

At last Armez yielded to another direct appeal of the emperor, and on December 15, 1866, the frontal part of the head of the illustrious Cardinal Richelieu was reinterred in its tomb in the chapel of the Sorbonne, after a discreet funeral ceremony presided over by the archbishop of Paris. First, however, the relic was subjected to an "anthropometric" examination, and a death mask was taken (a mere 224 years after death). Still later, in 1895, the tomb was opened yet again, this time for photographs of the face, and to satisfy the curiosity of the Minister of Foreign Affairs, who was a historian specializing in Cardinal Richelieu. Since then, this one small remnant of the man who was once the arbiter of European politics has been left in peace.

Emanuel Swedenborg:
A Price on His Head

This Swedish scientist, theologian, and mystic was a noted polymath of his time, making solid contributions to learning in fields as diverse as mathematics, mechanics, physics, chemistry, geology, navigation, geology, paleontology, and biology. And this was all in the first half of his long life, before he focused his attention on religion, mysticism, and the occult. He spent most of his eighty-four years in Sweden, but he happened to die on an extended trip to England, on March 29, 1772. He was buried in London's Swedish church, St. George of the East.

In 1908, the Swedish government obtained the body and reinterred it in Uppsala Cathedral. There was no head attached to the body, however, because, about fifty years after Swedenborg's death, a retired sea captain and amateur phrenologist took a fancy to it and stole it. (As we have seen elsewhere in this book, phrenologists were a pestilential nuisance in that era.) The skull changed hands several times in the next hundred years, and it finally surfaced in a Welsh antique shop. Some of Swedenborg's heirs learned about it and bought it.

The ancestral skull remained with the family until March 1978, when it unexpectedly appeared in an auction catalog of Sotheby's, in London. A Swedish bidder acquired it for $3,200. In this world, it sometimes takes money to get a head.

Francisco Goya: Unexpected Company

Goya was the preeminent Spanish painter of his time and one of the best artists the world has seen. Some of his works, such as the so-called *Naked Maja, Saturn Devouring One of His Children*, and *The Disasters of War*, are among the masterpieces of all time. Like Swedenborg, his contemporary, Goya spent almost his whole life in his native country, but he had the misfortune to die abroad. He expired on April 16, 1828, at eighty-two years of age, during a brief self-imposed exile in France. He was buried in the Cemetery of the Chartreuse in Bordeaux.

Years later, in 1899, Spain obtained permission to translate Goya's remains to Madrid for a splendid reburial in the Church of San Antonio de la Florida, beneath the very dome whose beautiful frescoes Goya had painted a century earlier. When the authorities opened his grave in Bordeaux, however, they had a problem. There were two skeletons inside, and they couldn't tell which one was Goya's. Even worse, there was only one skull, and no way to tell which pile of bones it belonged to. Mathematically, there was only a one-in-four chance that any given combination of skull and bones would be Goya, the whole Goya, and no one but Goya. So, rather than risk making a mistake or waste too much time on the problem, practicality prevailed and the whole lot of relics was carted off to Madrid and reinterred in a single magnificent sarcophagus marked *Francisco Goya.* Let God sort them out.

Phineas Gage Goes to Harvard

hineas Gage was just an ordinary nineteenth-century laborer. He never would have been admitted to prestigious
Harvard University while he was alive. For one thing, he
had very little formal education. He also had a hole in his head:
not a hole of the metaphorical kind, but an actual hole from
one side to the other, the result of a gruesome accident. However, it was just this hole in the head that got Gage into Harvard *after the funeral*, or at least, it got his head into a Harvard
museum. You can still see it there today, if you are not too
squeamish.

In September 1848, the twenty-five-year-old Gage was
foreman of a construction crew for the Rutland and Burlington
Railroad, laying track in Vermont. Part of his job was to drill
holes in the rock, pour in blasting powder, insert a fuse, fill
the hole with sand, then pound the mixture down tight with
a long metal tamping rod to prepare for blasting. One day he
made a mistake. He started pounding before the sand was
poured on top of the blasting powder. The powder exploded
and propelled the yard-long tamping iron at high velocity out
of the hole, out of Gage's hands, and right through his skull.

The thirteen-pound bar landed far away. Phineas fell
down, but soon stood up and walked and talked normally. He

was in terrible pain, and he was bleeding seriously from the entry wound in his left cheek and the exit wound in the top of his head. The tamping iron had smashed a path diagonally upward behind his left eye, one and a quarter inches in diameter, right through his brain.

His coworkers loaded the wounded man onto an oxcart and took him to a nearby tavern, where he was given a room. Someone thoughtfully picked up the tamping iron and brought it to him. Everyone thought he would die, but despite severe hemorrhaging and vomiting all night, he survived. He never lost consciousness, but he suffered intermittent delirium for a few weeks and lost the sight of his left eye. In two months he was able to work, but since the benevolent Rutland and Burlington Railroad had fired him, he went home.

Although he was as strong and intelligent as before the accident, Phineas was a changed man. Previously he had been sober, responsible, and socially well adjusted. Now he began to use profanity, lied, and couldn't be trusted. His physician, John Harlow, was intrigued by this personality change, and wrote two academic papers arguing that the injury to the frontal lobe of Gage's brain was the sole cause of it. The medical establishment scoffed at the naive suggestion of this country doctor that any specific region of the brain controlled one's personality and social behavior.

Soon, the patient began to roam around looking for work, and Dr. Harlow lost contact with him. Eventually he got as far as Chile, where he drove a stagecoach. But by 1861 he was back in the United States, ill and living with his mother and sister in San Francisco. He died the same year and was buried along with the fateful tamping rod, which he apparently had kept as a souvenir. There was no autopsy.

Meanwhile, Dr. Harlow continued his controversy with his medical colleagues, but he was getting the worst of it. His hypothesis that a limited area of the frontal lobe controlled social behavior was weakened by his inability to specify which area it was, since he had only been able to observe the exterior of poor Gage's head, not the interior of his brain.

Dr. Harlow finally learned of Phineas Gage's death about five years after it happened. He tried to get an autopsy report, but of course there was none. Finally, he asked the dead man's family to exhume Phineas and send him the head. This was an unusual request, to be sure, but the family complied with it. They also sent the tamping iron.

Dr. Harlow still was not able to convince his skeptical opponents, because the scientific knowledge and techniques of the day were not advanced enough to reconstruct, from the skull alone, which areas of the brain had been damaged. However, as he had foreseen, modern science has proved him to have been on the right track. The skull and tamping iron eventually found their way into the exhibition cases of Harvard University's Warren Anatomical Medical Museum, where they have remained ever since, along with a plaster bust taken from life of the unfortunate Mr. Gage. Using these artifacts as clues, Drs. Hanna and Antonio Damasio and others of the University of Iowa enlisted the aid of modern computer simulation techniques to determine the most probable flight path of the iron projectile through Gage's brain, thus localizing the injury with enough certainty to compare its effects on Gage with similar frontal lobe injuries that were medically well documented. Sure enough, Gage's symptoms were remarkably similar to those in comparable cases, and Gage's cranium, recovered and preserved through the foresight of Dr. John Harlow, is still contributing to science's gradual unraveling of the secrets of brain function.

PART II

HEARTS

A Word About Hearts

Hearts, like heads, frequently have received preferential treatment apart from the dead bodies to which they belonged. Hearts present more problems in this regard than heads, however. Whereas heads are relatively easy to detach, removing a heart can be messy and difficult. Also, except for size, human hearts look pretty much alike, thus making them unsuitable as proof of someone's death or as identifiable mementos. Hearts, lacking bones, are much harder to preserve than skulls. Rather than being even remotely decorative like skulls, hearts are repulsive and bloody masses of tissue. For these and other reasons, hearts have never been quite as popular as heads for separate treatment after their owners' deaths.

Nonetheless, in Western and other cultures, the heart is popularly conceived of as the very seat of life in men and women, and, even more importantly, as the repository of many virtues, especially love, bravery, compassion, and truth. Biologically, of course, this is nonsense, but a heart looks better on a St. Valentine's Day card than an adrenal gland, and *King Richard the Lionhearted* sounds much more imposing than *King Richard the Lion-livered*. Symbolically, then, the heart is extremely well entrenched in the popular psyche as the noblest of human organs, and therefore, despite the obvious drawbacks mentioned above to giving special honors (or dishonors) to a dead person's heart, there have been many attempts to do just that. Consider, for instance, the case of the poet Percy Bysshe Shelley.

How Shelley Gave His Heart to Mary

Percy Bysshe Shelley was born in Sussex in 1792. He and two friends, John Keats and George Gordon, Lord Byron, were the leading poets of the English Romantic movement. They all died young, within three years of one another. Keats expired first, in 1821, and was buried in the old section of the Protestant Cemetery in Rome. His death deeply affected Shelley, at whose invitation he had come to Italy, where he died.

Shelley commemorated him by writing the elegiac poem *Adonais*. In his introduction to this tribute, Shelley described the place where Keats's body reposed: "The cemetery is an open space among the ruins, covered in winter with violets and daisies. It might make me in love with death to think that one should be buried in so sweet a place." Prophetic words, these; for within two years, Shelley's own ashes were buried there, not once but twice.

His heart, however, was not among the ashes. Its crumbled remains were found thirty-three years later among the effects of his wife Mary, who kept them until she died, wrapped in silk and pressed inside her leather-bound copy of *Adonais*. This touching but macabre memento somehow does

Mary Shelley wrote *Frankenstein* and kept her husband's heart in her desk. (*Courtesy of Library of Congress*)

not seem out of character for Mary Wollstonecraft Shelley, who was, after all, the author of *Frankenstein*. The touching saga of Shelley's heart, which was rescued from oblivion by the disinterested labors of a devoted friend, unfolded as follows.

The Shelleys had been living in Italy since 1818 because their liberated lifestyle, Percy's atheistic proclivities, and the middling acceptance of his poems had made England uncongenial for them. Shelley wrote some of his best works in Italy, such as *Cenci* and *Prometheus Unbound*, despite such distractions as the deaths of two of his children, poor health, and his growing ennui. He even had nightmares and hallucinations, at one point imagining he saw Allegra, Lord Byron's deceased daughter, rising from the waves to beckon him. In light of his own imminent death at sea, this is significant, as is the fact that he could never get over the shock of his first wife's death by drowning.

On July 1, 1822, Shelley sailed from Lerici, where he had a villa, to Leghorn (Livorno), leaving Mary behind with their two-year-old son, Percy Florence. His objects in going were to lighten his melancholic spirits by a sail in his refurbished boat, the *Ariel* (formerly the *Don Juan*), and also to continue on to Pisa to confer with Lord Byron, the journalist Leigh Hunt, and others about a projected new magazine to be called the *Liberal*.

A week later, July 8, he was back in Leghorn. Both he and his sailing partners, Edward Williams and a cabin boy, Charles Vivian, were anxious to return to Lerici, fifty miles away across the Gulf of Spezia. Ignoring the signs of gathering afternoon thunderstorms over the hot, shimmering sea, Shelley and his companions set out under full sail in midafternoon and soon ran into a terrific squall. Disdaining to return to port with the local craft, or even to shorten sail and ride out the bad weather, they were last observed plowing imprudently ahead in the teeth of the gale. They were never seen alive again.

The *Ariel* apparently was driven under by the force of the wind, and its occupants drowned. As days passed and no bodies were found, that could have been the end of Shelley's story. Fortunately, however, six months earlier Shelley had met an energetic and adventurous seaman, Captain Edward John Trelawny, who had formed an instant and intense attachment to the young poet. He had been one of the last to see him alive as he sailed away from Leghorn, and now he tirelessly reconnoitered the coast until he learned that three men's bodies had been washed up on different parts of the coast ten days after the storm. After conducting the distraught Mary and her little son to Pisa to be with friends, Captain Trelawny set out in Byron's yacht *Bolivar* to identify the bodies, which, though badly decomposed, were recognizable from their clothes and effects.

Trelawny took charge of everything. He had to deal with the separate governments of Lucca, Florence, and Pisa, since Italy was not unified at that date and the corpses had drifted ashore in different countries. At all locations, however, strict quarantine regulations were in force, which required dead bodies washed in from the sea to be buried on the spot and covered with quicklime to hasten deterioration. Trelawny, after shelling out over £400 to construct suitable oak coffins for Shelley and Williams, was chagrined to find that the same quarantine laws, especially in Tuscany, blocked his plan to move the bodies in style to the English cemetery at Leghorn. They must be left where they were, to avoid the possible spread of "contagion."

To circumvent this restriction and rescue the deceased from the ignominy of obscure, dishonorable burial, Trelawny, Byron, and Hunt thought of an ingenious compromise that won the consent of the authorities. They would cremate Shelley and Williams on the beach, and their ashes, purified by fire of all imaginary contagion, could be carted away at the Englishmen's pleasure.

Trelawny, ever the man of action, naturally took charge of the preparation and execution of this novel but onerous scheme, with the intellectuals serving mostly as witnesses. First he needed a portable iron crematorium, which was specially constructed for him by a local blacksmith. He also commissioned two boxes of walnut wood covered with black silk velvet and embellished with brass plates for reception of the ashes. Finally, he had to organize two separate expeditions of friends, constabulary, health officials, and laborers bearing firewood and other necessaries.

All of this took several weeks, but on August 15 the remains of Edward Williams were duly unearthed near Migliano, reduced to ashes, and taken by Lord Byron in his carriage back to Pisa. The usually cynical Byron was unnerved by the five-hour operation. "Don't repeat this with me," he said to Trelawny, "let my carcase rot where it falls." The next day, August 16, 1822, Trelawny traveled by boat to Massa, near Via Reggio, and in the blazing heat began searching for poor Shelley's burial spot. Byron, Leigh Hunt, and the local officials arrived later, by land.

It took over an hour of digging to locate Shelley's grave in the sands, and the excavators accidentally cracked his skull in the process. Having finished the exhumation, they stoked the furnace and began the cremation. In conscious imitation of the ancient Greek observance, the mourners poured wine, salt, and oil on the body, as well as frankincense. In his recollections, Trelawny wrote, "Byron asked me to preserve the skull for him; but remembering that he had formerly used one as a drinking cup, I was determined that Shelley's should not be so profaned." Leigh Hunt somehow kept a fragment of the jawbone, which later was deposited in a marble urn at the Keats-Shelley Memorial in Rome. A few other tiny fragments of bone were kept for Mary. Trelawny, however, acting on impulse, obtained the best souvenir of all. According to Trelawny's memoirs:

> The corpse fell open and the heart was laid bare. . . . The fire was so fierce as to produce a white heat on the iron, and to reduce its contents to grey ashes. The only portions that were not consumed were some fragments of bones, the jaw, and the skull, but what surprised us all was that the heart remained entire. In snatching this relic from the fiery furnace my hand was severely burnt; and

had any one seen me do the act I should have been put into quarantine.

There has been more than a little skepticism that the heart would not have burnt to ashes, and that Trelawny could have plucked it out of the intense flames as he claimed without being detected. The official Italian records of the operation show that after lying buried with quicklime for a month, nothing much of the corpse was left to cremate except the skeleton. Forensic experts have speculated that if any internal organ survived both the quicklime and the flame, it was more likely to have been the charred liver than the heart. Yet, in a way, does it matter? Trelawny preserved *something* from Shelley's innards; of that there is no doubt. If he fancied that it was his friend's heart, or simply wanted others, such as Mary Shelley, to innocently believe that the dead poet's heart was miraculously preserved from destruction, we can be indulgent enough to go along with it. After all, these people were not called Romantics for nothing.

After the sad ceremony on the seashore, after putting Shelley's ashes in the walnut box and dousing the white-hot furnace in the ocean, the participants in the day's events dispersed. Captain Trelawny took the box of ashes on board the *Bolivar* and sailed to Leghorn. By now, Mary had decided to inter her husband's remains next to those of their son William, in the Protestant Cemetery at Rome. This apparently had been Shelley's own wish. Mary entrusted the task to the indispensable Trelawny. Unable to take the box of ashes himself, Trelawny forwarded it to the English consul in the Eternal City for safekeeping until he should arrive. The consul, Mr. Freeborn, obligingly stored the box in his wine cellar, but as Trelawny's appearance was delayed for over five months, to avoid trouble with the authorities he took matters into his own hands and interred Shelley's ashes with proper obsequies on January 21, 1823. It being forbidden to bury any more people in the overcrowded older part of the grounds where William Shelley was laid, the consul hoped to disinter the child and rebury him in the new section of the cemetery with his father. Unfortunately, an adult body was found under William's grave marker, and rather than disturb more graves looking for the body, Freeborn buried Shelley alone.

When Trelawny finally got to Rome in March, he was not satisfied with the grave's location. He had two tombs constructed near the pyramid of Cestius, at a niche in the angle of the ancient Roman walls, and over one he erected an enigmatic blank tombstone. In the other he reburied Shelley's ashes, and the tombstone was carved simply with his name, three of his favorite lines from Shakespeare, and the somewhat misleading Latin motto *cor cordium* (heart of hearts), for Shelley's heart was the one part of him that was not buried there.

Where was Shelley's charred heart? Leigh Hunt begged it from Trelawny, intending to keep it. Mary Shelley, not unnaturally feeling a sentimental attachment to it, as well as a superior claim upon it, importuned Hunt to relinquish it to her. At first he refused, but at the intervention of a mutual friend he grudgingly surrendered it. Mary ever afterward kept it with her or near her, and, as mentioned above, after she died in 1855, it was found in her desk, completely crumbled to powder. In 1889 it was interred with the body of her only surviving son, Percy Florence Shelley, in the vault he had built for himself and his mother in St. Peter's churchyard, Bournemouth, England.

Mary had wanted to be buried next to her husband in Rome, but the empty tomb Trelawny had constructed there was not intended for her. For years the blank tombstone baffled tourists, who began making pilgrimages to the graves of Keats and Shelley within a few years of their deaths. In late 1880, the eighty-eight-year-old Trelawny notified the cemetery caretaker that the empty tomb would soon have an occupant. An English lady arrived the following October to deliver a box containing Trelawny's own ashes for burial next to the bosom friend he had known for only six months, and who had been dead for nearly sixty years. The following inscription is now carved on the long mute tombstone:

These are two friends whose lives were undivided;
So let their memory be, now they have glided
Under the grave: Let not their bones be parted,
For their two hearts in life were single hearted.

Lord Byron:
The Unromantic Truth

Lord Byron, the paragon of the English Romantic movement who participated in the seaside cremation of his friend Percy Bysshe Shelley, was soon to tread the road of death himself. He died during his crusade to help liberate Greece from the Turkish Empire, and for his crucial financial aid, if not for his ineffectual military support, the Greeks were extravagantly grateful. His heart, like Shelley's, was removed from his body before burial. A story was long current that when Lord Byron's body was shipped back to England, his heart was left behind in Greece at the urgent request of the patriots he had tried so hard to help. In truth, that was not quite the way things happened. Here is the correct story.

Byron had been interested at a distance in the cause of Greek independence for some years during his sojourn in Italy. In July of 1823, he interrupted his prolific literary endeavors, including his masterwork, *Don Juan*, and sailed to Greece to participate personally in the war for Hellenic freedom. His first five months were spent at Cephalonia assessing the scene of operations, and his next three months at Missolonghi trying to whip a ragtag militia into shape, mediate the internecine quarrels of the insurgents, and arrange a large English loan for the provisional Greek government. He didn't live long enough to actually fight any Turks, however; beginning in February, his health collapsed.

He died on Easter Sunday, April 19, 1824, probably of virulent malaria exacerbated by epileptic seizures, exhaustion, and merciless bloodletting by his four doctors. He expired at dusk just as, in true Romantic fashion, a spectacular thunderstorm broke over the town.

His friends summoned a village woman to lay out the body, while the Governor-General of Western Greece proclaimed three weeks of mourning, suspended Easter festivities, and ordered a thirty-seven-gun salute. The torrential rains forced a postponement of the state funeral until April 22. In the meantime, the four doctors, attended by the editor of the *Greek Chronicle*, performed an autopsy. They found the brain swollen and inflamed, the heart enlarged, the liver diseased; also, for reasons unknown, they apparently amputated his deformed right foot, since it was discovered surgically detached when the body was inspected in 1938.

The autopsists did not replace the brain or other internal organs they had removed for analysis, but they put the skull back together again, disposed the corpse to look as presentable as possible, and called in the undertakers to embalm the body in alcohol. Unable to procure lead wherewith to make a leakproof coffin, the undertakers had to make do with an oblong packing case lined with tin. Draped in a black cloak surmounted by a sword and laurel wreath, this was the coffin in which Lord Byron attended his first funeral, at the church of St. Nicholas in Missolonghi. The body then was returned to Byron's house until it was decided what to do with it. The Greeks would gladly have given it a hero's burial, even as far as interring it on the Acropolis of Athens. Some of Byron's friends argued for this, including Captain Trelawny, who had arranged for Shelley's cremation and burial two years earlier.

Part of the trouble was that Lord Byron himself had expressed widely contradictory wishes about how his body should be dealt with, ranging from a desire to be buried in Poet's Corner of Westminster Abbey, "an honor which . . . I suppose they could not refuse me," to the remark he made several times that his mortal frame be left wherever he died, "not hacked or sent to England," and interred without pomp. The disagreements among Byron's retainers on this point were growing heated when the fortuitous arrival of the brig *Florida*, with the first installment of the English loan that Byron had arranged, precipitated a decision to repatriate the body to its home country on the return voyage of the same vessel.

The undertakers, meanwhile, had another try at embalming the body. This they did by securing the leaky packing case with iron hoops, drilling several holes in it, and sealing it inside a large barrel of alcohol. On May 3, the barrel was transferred to Zante for safekeeping, and it was embarked on the *Florida* a few weeks later. It arrived in England on June 29, 1824.

Lord Byron's death in the cause of Greek freedom was sensational news throughout England and the Continent. Many people crowded the vessel in its anchorage for a glimpse of the hero. Even the Marquis de Lafayette, who was passing through London at the time, begged permission to see the body privately. After the necessary customs formalities, the body was landed, identified, transferred from the barrel to a lead coffin, and given a public funeral after lying in state for seven days in London. Despite Byron's fame throughout Europe as a poet and freedom fighter, the dean of Westminster, who took a dim view of his scandalous private life, refused him a spot in the abbey; the dean of St. Paul's also refused to let him be buried in his cathedral. The heirs and executors, unable to have him buried as a great poet, were "forced to bury him as a lord"—at their own expense! He was accordingly taken in funeral procession from London, through Nottingham, to his family's burial vault in the village church of Hucknall Torkard.

But what about Lord Byron's heart? The romantic tale that it was buried in a temple in Greece is not true, but it has some foundation in misinterpreted facts. The Greeks of Missolonghi, bereft suddenly of their benefactor, had hoped to bury him in their native soil, but gracefully acceded to the final decision of his entourage to return the body to England. However, they pressed their case that perhaps one of the internal organs of Lord Byron, such as the heart, which had been removed at the autopsy and stored in separate jars, could be left behind in the land for which the poet had sacrificed his life. This wish was granted. An affidavit given by the undertakers when they had finished preserving the corpse explains what happened:

> We, the undersigned, bear witness that in the large case, which has been sealed with the seals of the Provisional Government, is to be found the authentic body of the Honourable Lord Byron, peer of England; that we ourselves placed the said body in the case, closed it hermetically, and thereupon affixed thereto the seals above mentioned.
>
> We bear witness also that in the smaller case will be found the honoured intestines of the said noble and respected Lord Byron . . . , that is to say, the brain, the heart, the liver, the spleen, the stomach, kidneys, etc., contained in four separate jars. The lungs, which are missing, were deposited, in deference to the repeated representations of the citizens of Missolonghi, in the church of San Spiridione, in the hope that the most noble and respected family of the Illustrious Lord would grant them to Missolonghi, of which town My Lord had accepted the honorary citizenship.

Thus the poet's lungs remain in a Greek church, but, alas, his heart does not repose in a Greek temple. The jars containing Byron's heart, brain, and other organs accompanied the body to England, and were deposited with it in the family vault in the church at Hucknall Torkard. They were still there when the vault was opened privately in 1938 to dispel rumors that the body had been stolen. The casket had indeed been broken into, and a few pieces of jewelry and brass pilfered, probably during church renovations about 1880, but the body, its heart tucked into a box beside the coffin, was still there.

Thomas Hardy:
The Return of the Native

Heart burial, that is, the separate interment of a dead person's heart, seems to have originated in Europe in the Middle Ages, as a natural consequence of the method of embalming sometimes used to preserve the bodies of royalty, Church prelates, and saints. This method, like that of the ancient Egyptians, necessitated removal of internal organs, including the heart. Sometimes hearts (and other organs as well) were carefully preserved, either with the body in a separate container or elsewhere. This fad gathered momentum among the upper crust of society, especially in France and England, and peaked during the so-called Romantic era in the eighteenth century, when it was occasionally applied to the hearts of poets and other great artistic personages, for sentimental purposes as much as for any practical reasons. The custom declined sharply thereafter. One of the last significant occurrences of heart burial was in the case of the English writer Thomas Hardy, who was born in 1840 and died in 1928.

Thomas Hardy, at the time of his death, was a widely acclaimed novelist and an indifferent, though prolific, poet. His ponderous, somber, and depressing novels, such as *The Return of the Native* and *Jude the Obscure*, were enormously popular in his day and have been inflicted on impressionable schoolchildren of English-speaking countries ever since. He was a native of the little town of Stinsford, in Dorset, and most of his novels

Thomas Hardy's ashes are in Westminster Abbey, but his heart is in Dorset. (*Courtesy of Library of Congress*)

are set in and around this area, which he loved. When his first wife, Emma, died in 1913, he buried her in Stinsford church-yard where his parents and grandparents were interred, and in his will he directed that he be buried under the same tomb-stone as she; he even had a place reserved for his name to be chiseled on the stone when the time came.

Despite this, some of his friends had other ideas. They thought he belonged to the nation as a whole and the world of letters, and that he deserved a better fate than to be buried in an obscure country plot. Although no literary person had been buried in Poets' Corner in Westminster Abbey since Ten-nyson in 1892, and no novelist since Dickens in 1870, they were determined that Thomas Hardy was worthy of such an honor. The prime movers in this enterprise were Sir James Barrie, author of *Peter Pan*, and Hardy's literary executor, Sir Sydney Cockerell. They must have been preparing their plans quietly even before the writer's death late on January 11, 1928, for when it happened they instantly sprang into action. Cock-erell opened Hardy's will the same night and somehow con-vinced himself that the funeral instructions did not absolutely require burial in Stinsford churchyard. Barrie had hurried by train to London even before Hardy drew his last breath, and early on January 12 was busy marshaling support for the Abbey burial from the prime minister, the editor of the *Times*, and other influential people. By the afternoon of that day, the dean of Westminster had given his consent. Barrie telephoned the news to Cockerell.

Florence Hardy, the poet's second wife, was reluctant to violate her husband's burial wishes, but she was too upset to resist the persuasions of Cockerell and Barrie. The other rela-tives, who were not even consulted, were displeased. The local people of Stinsford were not happy either to have the body of their favorite son thus snatched away by outsiders against his express desire, depriving Stinsford of the honor of burying it. A row was about to explode over the question when suddenly an unexpected compromise was reached. It is uncertain whose idea it was, but, with Solomonic wisdom, the factions agreed to divide the prize, burying Hardy's heart in Stinsford and his body in Westminster. This course of action possibly originated with Sir James Barrie, but more likely with the local vicar, the Reverend H. G. B. Cowley. The decision may have been suggested by an article on medieval heart burial that, by coincidence, had just arrived in the mail at Hardy's house. In any case, no time was lost. Hardy's doctor returned to the house on the evening of January 12 with a surgeon, who removed the heart, wrapped it in a towel, and secured it in a biscuit tin. The doctor took it home for safekeeping and returned the next day to seal it in a bronze urn, which was placed on the church altar for vigil.

In another shock to the family, the rest of Hardy's body had to be cremated, because Poet's Corner was getting too crowded to permit burial of any more full-size caskets. The officious Cockerell and Barrie took charge of Hardy's body on Friday the thirteenth, conveyed it to a nearby town for cremation, and delivered his ashes, in another bronze urn, to London.

A splendid funeral was held at Westminster on Monday, January 16, marred somewhat by heavy rains that soaked the overflow crowd outside the Abbey. Inside, where it was dry, the pallbearers included not only political bigwigs such as Ramsay MacDonald and Stanley Baldwin the prime minister, but the highest literocracy of the land, like Barrie, Rudyard Kipling, and George Bernard Shaw, while lesser dignitaries and VIPs filled the aisles. Simultaneously, down in Stinsford, a more modest ceremony was taking place to entomb Hardy's heart in his first wife's grave. Here, at least, the sun shone.

The actual circumstances of Thomas Hardy's heart burial were unusual enough, but they gave rise to a ludicrous anecdote that, though untrue, persists to this day in cheap books of "amazing facts." There are several variations, but in general

The grave of Thomas Hardy's heart and of his first wife, in Stinsford churchyard, Dorset. (*Photo by Edwin Murphy*)

the story is this: Hardy died in London, and his heart was mailed in a cookie tin to his sister in Dorset for local interment. When the poor woman opened the box on her kitchen table, however, a cat (or a dog) jumped up, grabbed the morsel in its mouth, and dashed into the woods with it, consuming it in whole or in part before being caught. The family, in embarrassment, went ahead with the burial anyway, but the box they buried was empty. The persistence of this legend, despite overwhelming evidence of its untruth, is truly amazing.

Richard the Lionhearted's
Disorganized Demise

R ichard Cœur de Lion is the epitome of the crusading knight: tall, handsome, noble, cunning, brave, and, as his opponents in battle had ample proof, an almost invincible fighting machine. Born in 1157 and king of England from 1189, he spent only a few months of his ten-year reign on English soil. Joining the Third Crusade soon after his coronation, he conquered Cyprus as a base of operations before proceeding to the Holy Land. His energy, prowess, and ferocity were largely responsible for the crusaders' capture of Acre from the Moslems, and his dogged advance toward Jerusalem against overwhelming odds was finally stopped by the departure of his French and Austrian allies more than by the armies of Saladin. Forced to make a truce, he departed for his own realm, only to be captured by his former ally, Duke Leopold of Austria, and held for ransom. When finally released in 1194, he returned briefly to England, but spent his last five years in France defending his continental possessions from the encroachments of the French king. He died of an infected arrow wound on April 6, 1199, at the siege of the castle Châlus-Chabrol.

Richard left explicit instructions about the disposition of

his corpse. His body was eviscerated for embalming and transported to the abbey church of Fontevraud in Normandy, there to be buried at the foot of the tomb of his father, Henry II. Old Henry probably would have shuddered and clung to life had he foreseen that the remains of his rebellious and belligerent son would be buried so close to his own. The abbey was ransacked during the French Revolution, and Richard's body was thought to be lost. However, it was rediscovered in another part of the church in 1910. History records that the resentful King Richard, never one to tolerate any opposition to his wishes, ordered that, as a posthumous insult, his entrails should be buried in Poitou, whose people had rebelled against him; however, they also are said with more plausibility to be buried where he died, at Châlus.

Richard left other burial instructions concerning his heart. This organ he granted to the citizens of Rouen as a reward for their loyalty to him. His exceptionally large heart was encased in a silver casket and interred with honor in Rouen Cathedral. Much later, in 1250, the casket was donated to help ransom St. Louis from the Saracens. The heart itself disappeared for centuries, but in 1838 a lead box was found hidden elsewhere in the cathedral with the inscription *Hic jacet cor Ricardi Regis Anglorum*. Inside it was indeed the famous Lion Heart, now "withered to the semblance of a faded leaf." Later the heart was placed in the museum at Rouen.

The English must have felt a little left out in the distribution of their popular king's organic remains. He'd barely set foot in his realm for ten years (this was probably one reason he was so popular, since absence makes the heart grow fonder), and now England was cheated of his body, his heart, even his intestines. While young, Richard had attended All Hallows Church in London, built a chapel for it, and expressed a wish that his heart be buried there when he died. At some unknown time during the next century, church officials began showing visitors a receptacle before the altar in this chapel, in which they claimed that Richard the Lionhearted's heart was enclosed. Edward I certainly believed the heart was there, and even obtained from the pope an indulgence for all those who contributed to the upkeep of the heart chapel. Although the claim was still being made until the twentieth century, few outside of England now believe it.

Robert the Bruce's
Posthumous Pilgrimage

Whereas Richard I took his lion heart on crusade and brought it safely home again, Robert I "the Bruce," king of Scotland, died before he had time to fulfill his crusading vow; but he tried to send his heart to Jerusalem after he died, by way of proxy. Whether this technicality would have satisfied his vow is a moot point, because the heart never made it any farther than Spain.

Robert the Bruce became king of Scotland in 1306. He is best remembered for having soundly defeated a large English army at Bannockburn in 1314. He died of leprosy after a twenty-three-year reign, most of it spent fighting the English. This left him no convenient time to go on crusade, but he always intended to go, and he never gave up easily on anything he wanted to achieve. As his malady rendered him ever more helpless, he summoned one of his retainers, Lord James Douglas. According to Froissart's *Chronicles*, before he died on June 7, 1329, Robert told Douglas, "Since my body cannot accomplish what my heart wishes, I will send my heart instead of my body to fulfill my vow; . . . take my heart from my body, and have it well embalmed . . . [and] you will then deposit your charge at the Holy Sepulchre of Our Lord."

When the king was dead, his followers, acceding to his wishes, removed and embalmed his heart. A document still exists in which Pope John XXII granted absolution to those who, in breach of Church edicts against mutilating or dividing the bodies of dead Christians, extracted Robert's heart for this pious purpose. The body was buried at Dunfermline Abbey; when exhumed during church alterations in 1819, the cracked sternum clearly showed that the heart had been removed.

The suitably prepared noble heart was delivered to the faithful Lord Douglas, who set out for the Holy Land in the spring of 1330. He carried the king's heart next to his own, in an enameled silver case hung around his neck.

Hume of Godscroft, the historian of the Douglas clan, asserts that Lord Douglas succeeded in his assignment and buried the royal heart in a gold box before the high altar of the Church of the Holy Sepulcher. But the evidence does not support this. First of all, an extant letter from King Robert to his son clearly states that Douglas was to lay his heart before the altar in Jerusalem, but rather than bury it there, to return with it to Scotland and inter it in Melrose Abbey. Second, poor Douglas never got to Jerusalem. He made the mistake of landing in Spain and, by way of getting a little advance practice in crusading, agreed to get involved in a war then in progress between King Alphonso of Castile and the Andalusian Saracens. At a battle near Cordoba, Lord Douglas and his men were killed. Douglas's body, and the king's heart, were packed up and sent back to Scotland. There, the heart was entombed, as intended, at Melrose Abbey. The box containing it was found during renovations in 1921. So, even though Robert the Bruce's clever attempt to carry out his crusader's vow after death was a failure, at least his heart was in the right place.

CHAPTER 15

Frédéric Chopin:
Home At Last

The celebrated Polish composer Frédéric Chopin spent most of his adult life away from his native land. When he died of pulmonary tuberculosis at Paris on October 17, 1849, he had not attained forty years of age; yet already his work had made him world-famous. His many nocturnes, études, polonaises, and concerti were greatly responsible for establishing the piano as a solo instrument.

Frédéric was widely and sincerely mourned when he died. His body was given a tomb in the cemetery of Père Lachaise in Paris, with a mediocre statue by a sculptor he detested and the curiously informal inscription *Fred Chopin*. But Fred's heart was always in Poland, and that's where he asked his doctor to send it after he had no further use for it. Accordingly, Dr. Jean Cruveilhier, who was with him when he died, removed his dead friend's heart. It was repatriated to Poland and immured with honor in the wall of the Church of the Holy Cross in Warsaw.

How a British Subject
Served Louis XIV

*L*e Roi Soleil, Louis XIV, was king of France from 1643 until he died, on September 1, 1715. During his long reign, he centralized almost all effective power in his own hands, at the expense of the nobles and the parliament. He also raised France and its monarchy to the pinnacle of prestige and glory. He practically invented the concept of absolute monarchy, and thus he was absolutely despised by the Jacobins of the French Revolution. Grave robbers raided his tomb at St. Denis and stole his embalmed heart. An English nobleman, Lord Harcourt, bought it. Later, he sold it to the dean of Westminster Cathedral, the scientifically minded Reverend William Buckland. It passed by inheritance to his equally scientific but decidedly eccentric son, Francis Buckland. Frank, as everyone called him, was a likable if unconventional scientist. His first love was fish, but all forms of anatomy and other natural sciences fascinated him.

Among his many enthusiasms, Frank Buckland was a founder of the Society for the Acclimatization of Animals in the United Kingdom, an organization whose goal was to increase the national food supply by importing and raising all kinds of exotic animals. As the society routinely feasted on buffalo, kangaroo, ostrich, and many other outlandish species, Buckland got into the habit of regarding anything organic as a possible meal, and the more unusual the better. He was known to have consumed delicacies ranging from sea slugs to garden slugs, from bluebottle flies to earwigs, from moles to porpoise heads, the gastronomic merits or demerits of which he reported in detail for the avid readers of his several periodicals, including *Land and Water*.

When you dined at Frank Buckland's house, you could never be sure what might turn up on your plate. Thus, it should come as no surprise what one startled visitor reported about one of these repasts. Frank told him: "I have eaten many strange things in my lifetime, but never before have I eaten the heart of a king." Buckland then calmly proceeded to consume the contents of his plate, which consisted of the heart of Louis XIV. He had withdrawn it from his immense collection of curiosities and put it to practical use. His collection, incidentally, also contained a lock of hair from Henry IV and the poet Ben Jonson's heel bone. As far as we know, Frank Buckland did not eat these.

Louis XIV had reigned longer than any other European monarch. He probably would have tried to hang on even longer had he known his heart was destined to become the supper of an irrepressibly unorthodox English gourmet.

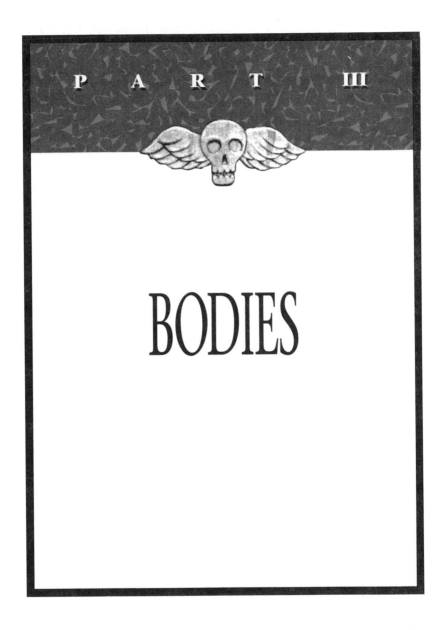

PART III

BODIES

A Word About Bodies

Having now considered some of the strange things that have been inflicted on the heads and hearts of prominent people after death, it is time we get into the body of this book and deal with bodies. In this part of the book, we shall see that almost anything can happen to a dead body, and most of the time the cause is the reverence, deference, cupidity, stupidity, curiosity, animosity, or other misguided motives of those still living, who hope to satisfy some of their own needs or further some of their own goals by utilizing the defenseless corpses of those who have gone before. There are exceptions, of course. Sometimes a not-yet-deceased person makes strange provisions for disposing of his own body. Not counting the fairly common, and laudable, custom of leaving one's body for scientific or medical study, or fads such as having one's body buried in a car or in the grave of a pet, there are some surprising cases of unusual self-inflicted indignities involving famous corpses.

For instance, the German poet Hans Wilhelm von Thummel insisted that he be buried inside a hollow oak tree in Noebdentz. The American stage magician and escape artist Harry Houdini left instructions to have himself buried in the trick coffin from which he used to escape during his act; as far as we know, he hasn't escaped this time. Richard Hull, an English

Jeremy Bentham's mummified body, dressed like this, sits in a corridor of University College, London. (*Courtesy of Library of Congress*)

eccentric, got his wish to be buried upside down on his horse, so that at the resurrection at the end of time, when "the world will be turned upside down," he will be in the right position to gallop away into glory. In California, that other magnet for eccentrics, "Harry the Horse," a member of the notorious Hell's Angels motorcycle club, arranged to have himself buried on his Harley-Davidson motorcycle: at the resurrection, he can gun the engine and roar off to—wherever.

More curious than any of these stories is that of the English economist and political philosopher Jeremy Bentham, founder of the notion of Utilitarianism. This creed held that everything must be put to the best possible use. In his pamphlet *Auto-Icon; or Farther Uses of the Dead to the Living*, Jeremy advocated that even dead bodies should be put to use, suggesting that everyone should become his own memorial by having his body preserved after death and exhibited in some convenient spot. To set an example, he stipulated in his will that his body be embalmed, varnished, dressed in his favorite clothes, and displayed in University College, London. This unusual request was honored after he died in 1832, and visitors to the college can still see Mr. Bentham ensconced inside a glass-fronted mahogany case set unceremoniously in a busy hallway. As a founder of the college, he was trundled to the annual board of directors meeting for ninety-two years, until the novelty wore off. Eventually his head, which didn't mummify well, had to be replaced by a wax effigy, but the real head is stored inside his rib cage in true utilitarian fashion.

Most of the really unusual things that have happened to famous dead bodies, however, were not of their owner's volition, but were instigated by other people. Following is a small collection of such stories, showing the vicissitudes that often befall innocent corpses *after the funeral.*

CHAPTER 18

Final Entrance: John Barrymore Bows Out

The Barrymores are easily the best known and most durable family of thespians in American history. For almost a century, there has seldom been a month when at least one of this numerous and talented dynasty of entertainers has not appeared on stage, on television, in the movies, or in the news.

John Barrymore, at one time considered the leading Shakespearean actor in the world, later went on to become a film idol. His ruggedly handsome face earned him the sobriquet "The Great Profile." Despite serious health problems in his declining years, he prided himself on never missing a performance. His most unusual appearance, however, was a brief cameo role, a decidedly low-key, nonspeaking part in a one-man show, presented in Errol Flynn's living room; it was unusual in that Barrymore was dead at the time!

There are three significant published versions of what happened at Flynn's house that night—four versions, if you count the one that says it never happened. However, since embarrassing or shocking events concerning famous bodies are routinely denied, disguised, or ignored by surviving relatives and adoring biographers, and are considered too irrelevant or disrespectful to find a place in standard reference works, we can discount denials of Barrymore's final curtain call and concentrate on the testimony of three witnesses to the event. Unfortunately, the three witnesses give widely discrepant variations of the relevant facts.

All variants at least agree on the main story line of this macabre episode: that someone borrowed John Barrymore's corpse from the funeral parlor, took it to Errol Flynn's living room, and set it up in a lifelike pose; and that Flynn, coming in late at night, was scared half to death when he saw it.

As for the plot details, however, everyone tells them differently. The glaring discrepancies among these renditions have been cited as evidence that nothing of the sort ever occurred. It is more likely, though, that the contradictions are attributable simply to so many instances of poor or convenient memory, careless research, gradual embellishment of a good story to make it better, or attempts to either steal the credit or avert the blame for an outrageous practical joke in undeniably poor taste.

Flynn tells the tale on himself in his autobiographical book *My Wicked, Wicked Ways*. Barrymore, destitute and dying, had been a cantankerous guest in Flynn's abode for three weeks before his final trip to the hospital, where he expired on May 29, 1942. According to Flynn, the cadaver was entrusted to the Pierce Brothers' mortuary on Sunset Boulevard. "I was particularly sad," wrote Errol. "I had come to know John pretty well in his declining days and had even felt favored by the three horrible weeks he had invaded my place."

Errol and some other friends, including the madcap director Raoul Walsh, gathered at a bar to commiserate on John's passing. Walsh, claiming he was too upset to socialize, pretended to go home. Instead, he and two friends (Bev Allen and Charles Miller) went to the funeral home and bribed the caretaker, giving him $200 to lend them Barrymore's body for a while. Transporting it to Flynn's house in Walsh's station wagon, they propped it in Errol's favorite living room chair, which they positioned to face the door. Flynn arrived late. He writes,

I walked in, sad and alone. As I opened the door I pressed the button. The lights went on and—I stared into the face of Barrymore! His eyes were closed. He looked puffed, white, bloodless. They hadn't embalmed him yet. I let out a delirious scream.

Errol bolted from the house, intending to flee in his car. His friends caught up with him on the porch and convinced him it was only a gag. "I went back in, still shaking. I retired to my room upstairs shaken and sober. My heart pounded. I couldn't sleep the rest of the night."

It is hard to imagine why Errol Flynn would publish such a story, which makes him the terrified butt of an unsavory practical joke, unless it had some basis in fact. He names Raoul Walsh as the perpetrator.

In his own book, *Each Man in His Time*, Raoul Walsh glee-fully takes full credit for the stunt, not deigning to mention any accomplices. According to Walsh, Flynn was not in a bar, but at home, drinking to Barrymore's memory, with only Walsh himself present. Errol excused himself temporarily for a visit to his lawyer's office to sign some papers. "I don't know what the hell made me do it," writes Walsh, who then tells of his trip to the mortuary of the Malloy Brothers (not the Pierce Brothers) on Temple Street (not Sunset Boulevard), where he asked Dick Malloy, a friend of his, to lend him the body. No mention is made of a bribe.

He then relates how he drove Barrymore's corpse to Flynn's house in his station wagon and enlisted the help of Flynn's Russian butler, Alex, to get the dead man into the house and onto his favorite spot on the sofa, with a drink next to him. Walsh then sat down facing Barrymore, waiting for Flynn to reappear.

> Errol came in, placed some papers on the table, and then saw Barrymore sitting on the couch. He let out a piercing scream, and ran out of the house. I went out to the doorway and saw Errol standing behind a big oleander bush. When he saw me, he yelled, "Get him out of the house, you crazy Irish bastard, before I have a heart attack."

Walsh repeats the same general scenario, with minor variations and additional details, in Richard Shickel's *The Men Who Made the Movies*. In its broad outline, Walsh's version of the event agrees fairly well with that of Errol Flynn. The third version, however, detailed in Paul Henreid's book *Ladies' Man*, while keeping the same plot as the others, substitutes a different cast in some of the roles. Barrymore is still the star and Flynn the costar in the farce, but Henreid enlists veteran character actor Peter Lorre, the quintessential Hollywood creep, as instigator of the morbid practical joke, and implicates Humphrey Bogart as well.

Henreid says that he, Bogart, and Lorre were shooting *Casablanca* the day John Barrymore died. Lorre conceived the prank and convinced the other two, with a couple of unnamed friends, to contribute several hundred dollars to bribe the mortuary. Although Henreid chipped in, he declined to help in the actual execution. He attributes the rest of his narrative to Lorre, who, he says, "giggled when he told me about it later."

Lorre knew Flynn would be out late shooting. He and the other conspirators smuggled Barrymore out of the funeral establishment and into Flynn's living room, arranged him in his favorite chair, and then hid behind doors. Finally Errol arrived:

> He opened the door and flicked on the lights and came in, threw his hat and coat on a chair and walked across

the room, past Barrymore's chair to the bar. He nodded at Barrymore and took about three steps, then froze. That moment was fantastic! There was a terrible silence, then he said, "Oh my God!" and he hurried back and touched Barrymore, then jumped. Barrymore was ice cold.

When Flynn realized it must be a joke, he shouted, "All right, you bastards, come on out!" He was angry, but also amused and trying not to laugh. He gave his friends a drink, but he refused to help them return the dead guest to the mortuary.

Countless variations on these versions of the Barrymore anecdote have appeared throughout the years in books, in articles, and on radio and television shows. The few deniers, who point to the factual conflicts between the different story lines as "proof" that nothing really happened, are themselves unable to agree on basic facts. Some blame Errol Flynn as the originator of the concocted tale, while others attribute it to Raoul Walsh. Whether true or not, the episode has become firmly enshrined in Hollywood lore, a part of the mythology of Tinseltown, where illusion and reality aren't all that different.

The great John Barrymore was buried on June 2, 1942, at Calvary Cemetery on Whittier Boulevard in East Los Angeles. The pallbearers included such notables as Louis Mayer, David Selznick, and W. C. Fields. As Barrymore's most famous role was Shakespeare's Hamlet, the tomb is inscribed with a line from that play, Horatio's farewell to Hamlet: *"Good Night, Sweet Prince."*

William the Conqueror's Funeral Fiasco

W illiam the Bastard is better known to history as William the Conqueror. In A.D. 1066, as every school-child is supposed to learn, the pugnacious Duke William conquered Harold's Anglo-Saxon army at Hastings and promptly promoted himself from duke of Normandy to king of England.

William the Conqueror was no one to trifle with. The original stormin' Norman, he was bold, strong, irascible, tough, authoritarian, and implacable—a real bastard. His whole life was one series of campaigns and battles, even after his conquest of England. He was detested, feared, and respected right to the end—and not a second thereafter. For his death and funeral must still rank as one of the most hilarious parades of undignified mishaps in royal annals.

Predictably enough, William met his death at the head of his troops, besieging the town of Mantes. The coup that felled the redoubtable warrior was somewhat ignominious: he ruptured himself on his saddle horn when his horse shied at a burning ruin. His retainers carried the stunned and dying king to the priory of Saint-Gervais in Rouen, where he expired some days later on September 9, 1087. Then the fun began.

The best description of William's funeral fiasco is from the pen of the respected Norman historian Ordericus Vitalis, who wrote a generation after the events. Scholars find little reason to doubt the accuracy of his account:

> The physicians and others who were present, who had watched the king all night while he slept, his repose neither broken by cries or groans, seeing him now expire so suddenly and unexpectedly, were much astonished, and became as men who had lost their wits. Notwithstanding, the wealthiest of them mounted their horses and departed in haste to secure their property. But the inferior attendants, observing that their masters had disappeared, laid hands on the arms, the plate, the robes, the linen, and all the royal furniture; and leaving the corpse almost naked on the floor of the house, they hastened away.
>
> Observe then, I pray you, my readers, how little trust can be placed in human fidelity. All these servants snatched up what they could of the royal effects, like so many kites, and took to their heels with their booty . . . Behold, this mighty prince, who was lately obsequiously obeyed by more than a hundred thousand men in arms, and at whose nod nations trembled, was now stripped by his own attendants, in a house which was not his own, and left on the bare ground from early morn to almost noon. . . .
>
> At length the religious, both clergy and monks, recovering their courage and the use of their senses, formed a procession; and, arrayed in their sacred vestments, with crosses and censers, went in due order to Saint-Gervais, where they commended the spirit of the departed king to God, according to the holy rites of the Christian faith. Then William, the archbishop, ordered the body to be conveyed to Caen, and interred there in the abbey of St.

Stephen the protomartyr, which the king himself had founded. His brother and other relations had already quitted the place, and all his servants had deserted him, as if he had been a barbarian; so that not one of the king's attendants was found to take care of his corpse. However, Herluin, a country knight, was induced by his natural goodness to undertake the charge of the funeral, for the love of God and the honor of his country. He therefore procured at his own expense persons to embalm and carry the body; and, hiring a hearse, he caused it to be carried to the port on the Seine; and, embarking it on board a vessel, conducted it by water and land to Caen.

Then Gilbert [de Coutances], the lord abbot, with the whole convent of monks, met the hearse in solemn procession, accompanied by a sorrowing multitude of clerics and laymen, offering prayers. But at this moment a sudden calamity filled the minds of all with alarm. For a fire broke out in one of the houses, and, shooting up prodigious volumes of flames, spread through a great part of the town of Caen, doing great damage. The crowds, both of clergy and laity, hastened with one accord to extinguish the fire, so that the monks were left alone to finish the service they had begun, and they brought the royal corpse into the abbey church, chanting psalms.

Afterwards, all the bishops and abbots of Normandy assembled to perform the obsequies of the illustrious duke, who was the father of his country. . . . After the mass, when the [empty] coffin had already been lowered into the grave, but the corpse was still on the bier, . . . Ascelin, son of Arthur, came forward from the crowd, and preferred the following complaint with a loud voice, in the hearing of all: "The land," he said, "on which you stand was the yard belonging to my father's house, which that man for whom you pray, when he was yet only Duke of Normandy, took forcible possession of, and in the teeth of all justice, by an exercise of tyrannical power, here founded this abbey. I therefore lay claim to this land, and openly demand its restitution, and in God's name I forbid the body of the spoiler being covered with earth which is my property, and buried in my inheritance." The bishops and other great men, on hearing this, and finding from inquiries among his neighbors that he spoke the truth, drew the man aside, and, instead of offering him any violence, appeased his resentment with gentle words and came to terms with him. . . .

However, when the corpse was lowered into the stone coffin, they were obliged to use some violence in forcing it in, because through the negligence of the masons it had been made too short, so that as the king was very corpulent, the bowels burst, and an intolerable stench affected the bystanders and the rest of the crowd. The smoke of incense and other aromatics ascended in clouds, but failed to purify the tainted atmosphere. The priests therefore hurried the conclusion of the funeral service and retired as soon as possible, in great alarm, to their respective abodes.

I have thus carefully investigated, and given a true account of all the manifestations of God's providence at the duke's death, not composing a well-feigned tragedy for the lucre of gain, nor a humorous comedy to provoke the laughter of parasites, but a true narrative of the various events for the perusal of studious readers. . . . A king once potent, and warlike, and the terror of the numberless inhabitants of many provinces, lay naked on the floor, deserted by those who owed him their birth, and those he had fed and enriched. He needed the money of a stranger for the cost of his funeral, and a coffin and bearers were provided, at the expense of an ordinary person, for him, who till then had been in the enjoyment of enormous wealth. He was carried to the church, amidst flaming houses, by trembling crowds, and a spot of freehold land was wanting for the grave of one whose princely sway had extended over so many cities, and towns, and villages. His corpulent stomach, fattened with so many delicacies, shamefully burst, to give a lesson, both to the prudent and the thoughtless, on what is the end of fleshly glory.

Nor was this yet the end of the indignities suffered by the Conqueror's corpse. In 1522 it was exhumed, examined, and reinterred; then in 1562 rampaging Calvinists rifled its tomb and scattered the bones, all of which were lost except one thigh, which was preserved and reburied in 1642. Alas, even this last relic was lost in 1793 during the riots of the French Revolution. It was rediscovered elsewhere in the church in 1988, and given yet another burial.

Sic transit gloria mundi.

C H A P T E R 2 0

The Queen Is Dead

Inez de Castro was crowned queen of Portugal in the year of our Lord 1360. She had been the mistress of the heir apparent to the throne for fifteen years. Her elevation to royalty didn't make the least impression on her; she neither changed expression nor moved a muscle during the entire ceremony. For, you see, in one of the most unusual royal coronations in history, Inez had already been dead for five years before the event.

On that hot summer's day in 1360, the crown descended on a head insensible of the honor bestowed, and the scepter was placed in the hand of one incapable of wielding it. The coronation of a corpse is surely one of the most outlandish caprices ever enacted in a civilized country. Besides showing that death is not necessarily an impediment to one's career, it also demonstrates once again that almost anything can happen *after the funeral.*

To understand why Inez was given her posthumous promotion from concubine to queen, it is necessary to know a little about the confused politics of medieval Portugal. Let's start with Afonso (or Alphonso) IV, king of Portugal from 1325 to 1357. His son and heir, Dom Pedro the infante, was born in 1320. When Pedro was sixteen years old, his father arranged for him to marry, by proxy, Constança, daughter of the powerful duke of Peñafiel in Galicia. War with Castile delayed the arrival of Dom Pedro's wife in Portugal for four years. Whether he ever loved Constança, who was chosen to be his wife for reasons of state, is unknown. What is certain is the violent passion he immediately conceived for her cousin and attendant, the ravishing Inez Pires de Castro.

Inez was a beauty. She had long yellow hair, and her natural elegance and grace earned her the epithet *Colo de Garça*, or Heron's Neck, which sounds much more complimentary in medieval Portuguese than in modern English. According to the vast literature their story inspired in every European language, the love between Dom Pedro and Inez was pure, sublime, and eternal. In reality, politics and dynastic intrigue played at least as large a role as love in the tragedy that followed.

For five years Dom Pedro distressed his wife and scandalized his austere father by an indiscreet liaison with Inez. Even though he had two sons by Constança, and even though Afonso exiled Inez to Albuquerque, Dom Pedro would not forsake his true love. When Constança died in 1345, Dom Pedro refused to take another wife if he could not have Inez. For various reasons, this was impossible. Inez eventually was allowed to return to court but only as a concubine. She bore Pedro four illegitimate children in the next ten years.

It was bad enough that Inez and Pedro were engendering a bastard line of potentially troublesome claimants to the throne. This was a development rife with future instability for a country where the royal succession was frequently tumultuous. But Inez's family, the Castros, were ambitious and growing in power, and Dom Pedro was falling under their influence. Intervening in both Portuguese and Castilian politics, the Castro family aroused increasing resentment and opposition from their rivals in the two countries.

The last straw came when Dom Pedro, in 1354, claimed the throne of Castile. King Afonso, not wishing another war with this powerful kingdom, managed to nip that scheme in the bud, but clearly things were getting out of hand. Now some of the royal councillors convinced Afonso that sooner or later there would be an attempt to supplant his royal line on the throne by elevating one of Inez's bastards. Something must be done.

Unfortunately for Portugal, though fortunately for world literature, which was subsequently enriched by hundreds of poems, plays, novels, and operas on the subject, the king decided that the best thing to do was to kill the beautiful but inconvenient Inez. On January 7, 1355, the chief justice of the kingdom and two other advisers persuaded Afonso to ride to Coimbra, where the infante and Inez resided, and put an end to her while Dom Pedro was absent on a hunt.

Afonso lost his resolve for the murderous deed when confronted by the beauty and tears of his intended victim, and by the entreaties of her children. He left the palace, but his three advisers returned almost immediately, perhaps with his tacit consent, and savagely slew Inez with their swords. Such, at least, is the legend, although some modern historians believe that Inez was executed with at least a semblance of legality, and that the three advisers were her judges, not her assassins. She was buried immediately with little or no ceremony in the nearby church of Santa Clara.

This drastic solution to Afonso's political dilemma completely backfired. The grief-stricken Dom Pedro, with military support from the outraged brothers of Inez, openly raised the standard of revolt. The short but bloody civil war that ensued was only settled by the mediation of Queen Beatriz, Dom Pedro's mother. Under the terms of settlement later in 1355, the infante received the post of chief justice of the realm and was given a large share in the government. Afonso's three ill-starred advisers, having botched things rather badly, fled for their lives.

Alas, two of them didn't flee far enough, for when Afonso

died two years later, one of Dom Pedro's first acts after becoming King Pedro I was to have them extradited from Castile and publicly executed in his presence. Pedro still mourned for Inez, and still refused to take another wife. He also wished, as everyone had long suspected, to make one of his bastard sons the heir to the throne of Portugal, ousting his own legitimate sons by Constança.

He met with much opposition to his dynastic plans, not only from the Portuguese nobility, but also from the Church, which could not sanction such a scandal, and from the nervous Castilian monarch, who didn't want a descendant of the Castro family occupying a royal throne just next door. Pedro had to resort to all kinds of stratagems and subterfuges to get what he wanted.

There were two major obstacles, among many others. The first was that, having never married Inez, he could not legitimize her children; second, since the mother, though noble, was not royal, her children could never legally inherit the throne ahead of Pedro's legitimate sons by Constança, who was both legally married and royal.

To solve the first difficulty, that of legitimacy, King Pedro claimed that, with special papal dispensation, he had secretly married Inez at Braganza in 1354. This was probably a lie, and certainly he had trouble proving it. Few believed it, but out of fear or self-interest many pretended to believe it.

To solve the second difficulty, that of rank, was more of a problem. A clandestine marriage to the infante, even if true, didn't necessarily confer royal rank on Inez. It was in this context that Pedro supposedly hatched the novel expedient of a posthumous coronation of his dead sweetheart's corpse. Whether it really happened or not, this unexampled quirk earned Pedro a footnote in history as one of the world's great royal crackpots. It also earned him an indelible place in literature and art as the truest, most ardent, most faithful of lovers. Antonio Ferreira's *Castro*, the first Portuguese tragedy, was inspired by the theme of Pedro's unparalleled constancy, as was a famous passage in Camöens' *Lusiads* and literally hundreds of other major and minor works in every European language for the next five hundred years. Even today it is a well-known romantic motif in Portuguese- and Spanish-speaking countries.

In fact, most of the juicy details about the macabre event are from literary, not historical sources, and the writers in this long literary tradition were not immune to Murphy's first law of storytelling: *However meager its basis in fact, a good tale gets better with every telling.* Since the only documentary evidence for this tale survives in brief extracts from obscure medieval Portuguese chroniclers, unavailable or unintelligible to most researchers, the writers and even the historians of each generation have tended to retell the story of the "Queen After Death" by recourse only to the plays, poems, and ballads of the previous generation. Each age adds new dramatic incidents for literary effect, which the succeeding age accepts as facts upon which, in turn, they pile their own improvements.

Now the literary tradition, in its bare essentials, tells that King Pedro I, whether from undying love or grief-induced insanity, in order to vindicate the honor of his dead "wife" and force her detractors to recognize her as the rightful queen of Portugal, exhumed her corpse, translated it with magnificent pomp to the royal court, arrayed it in gorgeous attire, seated it on the royal throne beside himself, and after a regular ceremony of crowning and anointing, obliged the nobles to come forward one by one to kiss its hand and pay their obeisance to Inez as sovereign queen. This story is told in many different ways, often at great and circumstantial length, and usually in strains of romance, drama, and bathos.

Is this what really happened? How much of this literary version of events actually took place? The slender evidence of contemporary writers gives many of today's historians cause for doubt. The main source is Fernão Lopez, royal archivist of

Portugal from 1418 to 1454, one of the most accurate, detailed, and reliable of medieval chroniclers. His *Cronica de el-Rei D. Pedro I*, which has never been fully translated into English, has this to say in chapter forty-four:

> And having thought to honor her bones, as there was nothing else he could do, he ordered a sepulcher carved of pale stone, all very cunningly executed, placing her image upon the upper stone, with a crown upon her head, as if she were queen. And he ordered the sepulcher placed in the monastery of Alcobaça, not at the entrance, where rest the kings, but inside the church, on the right, close to the principal chapel.
>
> And he had her corpse brought from the monastery of Santa Clara de Coimbra, where she had been laid, in the most exalted procession that could be arranged. She came in stages, a procession with extremely correct protocol for the time, carried by great cavaliers, accompanied by gentlemen of noble birth and many other people, and ladies, and damsels and a great number of clergy.
>
> By the sides of the road stood many men with great candles in their hands, organized in such a way that, wherever the corpse went, along the entire route, it travelled between lit candles. And thus they arrived at the aforesaid monastery, seventeen leagues distant, where with many masses and great solemnity, her corpse was placed in that monument. And that was the grandest funeral procession which had been seen in Portugal as of that time.

Fernão Lopez wrote more than sixty years after the event. His failure to relate most of the interesting details which later embellished the poetic version of the coronation, such as nobles kissing the dead and shriveled hand, is not necessarily proof that nothing of the sort ever happened. After all, absence of evidence is not evidence of absence. However, given the fact that Fernão Lopez was a meticulously detailed chronicler with a decided flair for relating colorful and dramatic incidents which his extensive research unearthed from royal, civic, and ecclesiastical archives, it is at least significant that he did not give a picturesque account of something as singular as the actual coronation ceremony, with all the trimmings, of a dead woman. Perhaps the mention of the crowned image of Inez on the lid of her sarcophagus was misinterpreted by later readers as a crown being placed on the actual corpse's head.

It is undeniable that Pedro I did try to obtain posthumous recognition of Inez both as his legitimate wife and as the legitimate queen of the realm. He undoubtedly had her body disinterred from its plebeian plot in the convent of Santa Clara in Coimbra, brought it by regal procession to the monastery of Alcobaça, as Fernão Lopez relates, and buried it there in state with members of the royal family. He built the twin sarcophagi at Alcobaça, whose exquisite beauty and consummate workmanship were absolutely unique for their time and place. They still exist today. Inez was reinterred in one of these twin monuments about 1363, and King Pedro in the other at his death in 1367.

There the two lovers rested in peace for a long time, while writers, musicians, artists, and poets spun their story of passion, intrigue, murder, revenge, and vindication into an ever more elaborate legend. The fame of Inez's beauty was so widespread that in 1810 or 1811 it led to one final disturbance of her cadaver.

At that time, during the War in the Peninsula, Napoleon's invading army arrived at the monastery of Alcobaça. Some undisciplined soldiers from Count Erlon's division under General Masséna broke open her beautiful tomb in search of loot. They had to settle for a souvenir. Handsome yellow hair still adorned the mummified head of Inez, and the vandals cut much of it off for keepsakes. What became of it is not known, except for one lock which was reportedly sent as a present from France to the emperor of Brazil but was lost at sea around 1846.

The beautiful Inez de Castro has not, so far, been disturbed again. Her body, having been unearthed first for a coronation and later for a haircut, resides peacefully in the exquisite portrait sarcophagus made for her by her dear Dom Pedro, who rests beside her.

Incidentally, all Pedro's scheming to put a son of Inez on the throne came to naught. When Pedro died, he was succeeded by Ferdinand, the legitimate heir born to his royal wife Constança.

CHAPTER 21

Will the Real Columbus Please Stand Up?

In 1992, the world celebrated the five-hundredth anniversary of Christopher Columbus's discovery of America. But if, in honor of this event, you had planned a pilgrimage to lay a wreath on Columbus's grave, you would have had a problem. It seems that no one knows for certain where Christopher Columbus is buried!

It's not that his body has been lost; rather, it is a case of too many bodies in too many places. Columbus traveled a lot during his life, and he didn't stop when he died. His corpse was moved so often during the last five centuries that we no longer are sure exactly where it ended up. At last count, four cities claim to possess the great explorer's authentic remains, and the dispute over who has the right cadaver has never been resolved.

Columbus's bones have become bones of contention, coveted by cities anxious for the renown (and tourist potential) of being their final place of rest. The saga of Columbus's corpse, its many wanderings, and its disputed location in the twentieth century is a fascinating tale of neglect, confusion, and mixups on a grand scale.

Such controversy over Columbus's body is an appropriate

The altar of the old cathedral, Santo Domingo. Christopher Columbus is buried here—and in three other places as well. (*Courtesy of Library of Congress*)

postscript to its owner's stormy career, for controversy surrounds every aspect of his life. His date and place of birth are not certain; the chronology of his early years is much disputed. It is unclear when or what he studied at the University of Pavia, whether he visited Iceland, and where he obtained his revolutionary geographical ideas. Even the nature of his discoveries was long disputed, Columbus himself stubbornly insisting he had reached the eastern fringe of Asia. In some quarters, Columbus was even denied credit for being the first to cross the Atlantic, and for many years his discoveries were erroneously attributed to Amerigo Vespucci. That is why two continents are named America, not Columbia.

Late in life, Columbus was accused of misgoverning the Spanish colonies he had planted in Hispaniola. He fell into disgrace, lost favor with the Spanish Crown, and spent his remaining years unsuccessfully prosecuting claims for his lost wealth and titles. This protracted litigation against the king made him even more unpopular. At last, worn out by illness, disillusionment, and a life of hardships, he sank into the arms of Death.

Most sources agree that Columbus expired on May 20, 1506, at Valladolid, an inland town of northern Spain. Though he was neither poor nor alone, as later legend claimed, the discoverer of America died in profound obscurity. Since he was in disgrace with the king, Columbus was socially a nonperson and his passing was barely noticed. The official chronicle of Valladolid omitted him from its daily register of the deaths of prominent people, and only several weeks later did a town document note laconically that "the said Admiral is dead." New editions of Columbus's accounts of his voyages, published up to three years later, showed no awareness of the author's death.

Having died unnoticed by a world that cared not, Columbus received a humble burial at Valladolid. It was so humble, in fact, that there is no conclusive evidence of where he was interred, and from this inauspicious beginning grew the chain of misadventures, still unresolved, which seems to have translated the poor admiral's body to four places at once. Tradition holds that his first resting place was in the church, or perhaps the crypt (sources vary), of a Franciscan monastery in the town. The monastery no longer exists, and lately the site has been occupied by the billiard rooms of the Cafe del Norte.

Between 1506 and 1514 (again, sources disagree), the explorer's body was transferred to Seville and reburied at the Carthusian monastery of Las Cuevas. Predictably, sources are at odds whether the burial took place in the choir, the vault, or the chapel. So far, the record is confusing enough, but soon it becomes even more perplexing. Several royal orders were issued, in response to petitions by Columbus's heirs, to accommodate the admiral's reputed wish to be buried in Hispaniola. On June 2, 1537, August 22, 1539, and again on November 5, 1540, the Crown ordered the removal of Columbus's remains to Santo Domingo. Strangely, however, records in the monastery of Las Cuevas show that the body already had been surrendered for transport to the New World in 1536.

The trail of Columbus's bones grows really murky at this point. Apparently, his remains were indeed transferred to Santo Domingo sometime between 1536 and 1549, though documentary evidence is contradictory. The year usually speculated for Columbus's burial in the Cathedral of Santo Domingo is 1541, soon after the completion of that edifice. Recent research indicates that the bishop initially refused him interment in the mortuary chapel of the high altar, as his rank demanded, because he was "a layman and a foreigner," and also, perhaps, because his acrimonious term as governor decades earlier had neither been forgotten nor forgiven. This episcopal recalcitrance may have necessitated the repeated royal orders mentioned above. Meanwhile the body, tucked in a leaden casket, is said to have moldered in an underground chamber beneath the cathedral until the dispute was resolved, although the real sequence of events is obscure.

In fact, there is no contemporary record of where in the cathedral Columbus's body was finally deposited. Not until 1676 was an entry made in church documents that Columbus was buried to the right of the altar, and in 1683 the memories of old people were cited in substantiation of this placement. Nevertheless, an opposite tradition somehow arose that Columbus was buried on the left side of the altar, not the right. About a century later, in the course of some repairs, stone burial vaults were discovered on both sides of the altar. In accord with tradition, but contrary to the earliest known (though not necessarily accurate) record, the sarcophagus on the left was judged to be that of Christopher Columbus; the other was believed to contain the remains of his brother Bartholomew, which along with those of his son Diego and grandsons Don Luis and Christopher II had all eventually found their way to the Santo Domingo cathedral.

When the eastern part of Hispaniola was ceded temporarily to France by the treaty of Basle in 1795, neither the Spanish authorities nor Columbus's descendants wanted the admiral's body left behind in French possession. On December 20 of that year, the governor, clergy, and people assembled at the cathedral for a solemn disinterment. Inside the left-hand vault they found fragments of a lead coffin, some bones, and the disintegrated mold of a human body, all of which were carefully gathered into a gilded casket and borne with ceremony to a ship in the harbor. With majestic pomp, the remains of Columbus were carried to Cuba and reburied in the cathedral of Havana.

Or were they? No one questioned the official identification of the body parts at the time, but in 1877 workmen at the cathedral in Santo Domingo again opened the two burial vaults. According to inscriptions, the one on the right contained the bones of Don Luis, the grandson. The left-hand vault, being empty, presumably was that from which Columbus had been removed in 1795. But the workers also discovered a hitherto unknown burial chamber hidden behind the empty vault. This secret chamber immediately threw all prior assumptions into doubt, for a leaden casket found within it contained bone fragments, mold, a lead bullet, and a silver plate reading, *a part of the remains of the first Admiral, Don Christopher Columbus!*

Another inscription on the casket itself reads, *Illustrious and renowned man, Christopher Columbus.* The initials C.C.A. on the sides of the box probably stand for *Cristobal Colon Almirante* (Christopher Columbus, Admiral). But if the remnants of Columbus's mortal frame were still within this box in Santo Domingo, then whose had been relocated to Havana many years before? Most likely they were those of Don Diego, Columbus's son, who also had been buried somewhere in the sanctuary near his father; brother Bartholomew or grandson Christopher II were also possibilities, since both were buried in the cathedral as well.

At first, the newly discovered remains were guardedly accepted as genuine; but an official Spanish inquiry by the historian Manuel Colmeiro declared that Columbus's body had indeed been moved to Havana in 1795 and that the remains lately coming to light in Santo Domingo were those of Don Diego. The Spanish Academy of History at Madrid agreed that Columbus made it to Havana, although it concluded that the body in the C.C.A. casket belonged to grandson Christopher, not Don Diego. Imputations were leveled that the inscriptions identifying the occupant of the C.C.A. casket as Columbus were fraudulent additions or alterations.

Undaunted, the ecclesiastical authorities of Santo Domingo launched their own inquiries and commissioned their own experts to prove that Don Diego had been shipped to Havana by mistake while Columbus himself was left behind.

Rumors were resurrected (or invented) that an error of this sort had been suspected in Santo Domingo all along. It would have been easy for a mixup to occur, even as far back as 1585. In that year Sir Francis Drake threatened to burn the town, and it was alleged that the priests had hurriedly exhumed both bodies for safekeeping; they may not have put them back in their proper positions. Besides, alterations had been made in the great chapel at various times; the altar itself had changed position more than once. Anything might have happened to the bodies during all these years! In any case, to share their luck at having rediscovered the discoverer of America, and also, incidentally, to win adherents for their point of view, the local authorities solemnly presented small relics of the admiral's bones to the Vatican, to the city of Genoa, and to the University of Pavia, all of which hold them in suitable honor to this day.

The debate between Santo Domingo and Havana simmered for twenty-one years before events took a new turn. In 1898, when Cuba obtained independence, the Spanish authorities moved the disputed relics once again, from Havana to Seville, where they are now displayed on a catafalque in (where else?) the cathedral. As a goodwill gesture, the Spanish government dispatched a bit of this precious dust to Genoa, where, unlike the relics on display from Santo Domingo, it is kept locked in a vault at city hall. Apparently the Genoese still support the claims of Santo Domingo, but they are in a position to switch allegiance at a moment's notice.

So the two main claimants to Columbus's body are Santo Domingo and Seville. But there are other rivals for the honor of being the final resting place of the famed explorer. The city of Havana has some evidence that the bones and ashes shipped back to Spain in 1898 were not necessarily those received from Santo Domingo in 1795. Few outside Cuba give serious consideration to this claim, but Eusebio Leal Spengler, city historian of Havana, prepared and circulated a research paper supporting the Havana hypothesis.

Other scholars argue, somewhat weakly, that Columbus's corpse never left Seville for transportation to the New World in the first place, thus giving Seville two claims on the body instead of one. However, the most intriguing theory so far, advanced by some Andalusian historians in 1949, holds that Columbus's body never stirred from its original resting place in Valladolid. In a practice common in his day, Christopher Columbus became a monk on his deathbed. Having joined the Franciscan order, he naturally was interred in the Franciscan monastery. According to the proponents of this scenario, the Franciscans never would have surrendered one of their own brothers for reburial at a distant town in the convent of a rival order, the Carthusians. If, as this unproven assertion has it, the monks delivered an impostor's corpse for removal to Seville, the body of Christopher Columbus, Admiral of the Ocean, must still rest in the monastery crypt, below the billiard rooms of the modern Cafe del Norte!

Thus Columbus's body is claimed to lie in five different spots in four different cities: Santo Domingo, Seville, Havana, and Valladolid. Reputed relics of his body are kept at the Vatican, at Genoa, and at the University of Pavia. There are other strands in the tangled skein of controversy surrounding the bones of Columbus, and the volume of literature on the subject is truly astonishing. An attempt was made to sell the Santo Domingo remains for display at the World Columbian Exposition in Chicago in 1892. About the same time, small quantities of these ashes were obtained by John Boyd Thatcher, mayor of Albany, New York, and by others. At various times, these souvenirs were displayed in the Library of Congress, the Lennox Library in New York, and sundry private homes. Little pinches of them have been worn in crystal lockets, handed down as family heirlooms, and kept in bank vaults from New

York to Hollywood as treasured personal possessions. Nor is the saga over yet. The government of the Dominican Republic moved its Columbus once more in celebration of the quinquecentennial of the discovery of America in 1492. An enormous lighthouse was constructed in Santo Domingo as the new resting place of the revered relics, a move that rekindled the old dispute among the rival cities over whose Columbus is the real McCoy.

But where does all this leave us? Rather, where does it leave poor Columbus? Unless new evidence comes to light, the mystery will remain unsolved. The most scientific and impartial investigation so far, a forensic analysis of all bone fragments in Seville and Santo Domingo in 1959 and 1960, concluded that parts of two distinct bodies reside in each location, with no duplication but many parts missing altogether. They are probably the remains of Columbus and his son Diego, but there is no actual proof. A proposal by an American anthropologist to settle the affair by measuring strontium ratios in tooth fragments, using mass spectrometry, was blocked by political reluctance and bureaucratic inertia.

The case of Columbus's corpse remains an open controversy. It is a fantastic, but by no means unique, lesson in the strange odysseys that can befall the bodies of even the world's most celebrated people *after the funeral.*

Where's Voltaire?:
The Scattered Philosopher

Voltaire, the renowned French writer and political philosopher, led a most varied and interesting life. *After the funeral*, his body continued to have interesting adventures of its own—and most of them shouldn't have happened to a dog. His heart and brain went their separate ways, other parts of his anatomy were snatched away at various times, and the remainder was stolen and spitefully destroyed by vandals. Only the heart remains now, locked in the base of Voltaire's statue in the Bibliothèque Nationale in Paris. How the corpse of such a rich and famous man came to such a terrible end is an interesting study in the unpredictability of human affairs.

Voltaire, whose real name was François Marie Arouet, was the embodiment of the so-called Age of Enlightenment of the eighteenth century. His prodigious literary output, encompassing poetry, drama, history, essays, satire, science, and philosophy, almost always had a political and social purpose, championing the freedom of the individual against abuses of power by the established authority. Not surprisingly, the established authorities of his native France generally looked askance at this vocal critic, whose writings became more popular the more they were banned, and who grew more influential the more he was persecuted.

Still, Voltaire eventually had endured enough persecution, in the form of imprisonment and exile, that he no longer relished that part of his job as reformer. Besides, he had grown famous throughout Europe, and quite rich. Finally, in 1760, at age sixty-six, he settled in Ferney, just inside the French border with Switzerland. There, he was poised for a quick exit across the frontier in case his enemies made things too hot for him in France.

In 1778, at eighty-four years of age, Voltaire made an ill-advised trip to Paris, where he was received with much popular acclaim and much official coldness. But the exertion was too much for his age and poor health. He died in great pain on the night of May 30 of that year, at a house owned by his friend the marquis de Villette. Crowds of admirers and well-wishers congregated outside his lodging unaware of his demise, clamoring for him to show himself on the balcony. However, the prime concern of his friends and relatives inside was to conceal his death from the authorities. Since Voltaire's anticlerical and skeptical writings had earned him many enemies in the Church and state, and since his lukewarm deathbed confession and recantation had not been satisfactorily completed, he died without receiving the last rites.

Thus, even though he had not been formally excommunicated, Voltaire still would not be permitted a Christian burial by the Paris authorities. His lifelong fear that his body would be cast on a rubbish pile seemed about to become a reality. His two nephews, Abbé Mignot and Monsieur d'Hornoy, who were members of the Paris Parlement, had tried desperately to get high officials, including the king himself, to intervene with the archbishop of Paris and permit Voltaire a Church burial when he died. All was in vain. The best they could secure was permission to transport the philosopher's corpse, when the time came, to Ferney, where Voltaire had prepared his own tomb. This permission was given without the knowledge of the archbishop, who did not yet know that Voltaire had expired.

While these negotiations went on, surgeons and an apothecary autopsied and prepared the corpse. Monsieur Mithouart, the apothecary, asked the others if he could keep the philosopher's brain; no one objecting, he boiled it in alcohol and took it home in a jam pot. The marquis de Villette, Voltaire's friend and host, claimed the heart, putting it somewhat more fittingly in a silver box plated with gold. The residue of the deceased was sewn up, the trepanned skull held together by a tight cap.

Speed was essential. The nephews wanted to get Voltaire's cadaver out of town and safely buried in a church before the news leaked out. Also, after two days, the smell of the corpse was not getting any better. Voltaire's body was dressed in a gown, bandaged to maintain a seated position as though alive, and held upright in a coach by cushions and concealed straps. This charade fooled everyone, and soon the coach was speeding through the countryside, not to Ferney, where a church burial could still be refused or reversed under orders from Paris, but to the Abbey of Scellières (or Seillières), near Troyes in Champagne, where Voltaire's nephew, Mignot, was commendatory abbot and could order the prior about as he pleased.

The body arrived at the abbey at noon the next day, where it was straightened out, placed inside a quickly made pine box, and buried the following morning under a flagstone in the church. Catholic rites and six masses were performed over the fugitive philosopher before ecclesiastical orders were received on June 3 forbidding the prior of Scellières to bury Voltaire in his abbey. When the archbishop of Paris learned that Voltaire already had been interred, he caused the bishop of Troyes to order the prior to exhume the body. The prior and Voltaire's nephews objected that this was impossible, asserting falsely that the cadaver had been sunk under two feet of corrosive quicklime. The bishop therefore contented himself with dismissing the prior, and the archbishop comforted himself with the thought that at least he was rid of the impious Voltaire for good.

The authorities forbade any mention in the press of Voltaire's death, any commemorative religious services, and any kind of monument to him anywhere in France. The nephews instituted unsuccessful lawsuits to recover Voltaire's brain and heart, which had been removed without their knowledge. The years passed, and while Voltaire's body disintegrated slowly in its grave at Scellières, the social fabric of the country disintegrated rapidly, culminating in the French Revolution, which exploded in July 1789.

The marquis de Villette became an enthusiastic republican, renouncing his title to become plain Citizen Charles Villette. He witnessed a surge in popularity of Voltaire's plays and books, no longer repressed by government authority. In fact, Voltaire's fame with the masses soared to new heights. The Abbey of Scellières, however, having been seized by the state, was gone to ruin and about to be demolished and the ground sold. Voltaire's bones again were on the verge of the dreaded rubbish heap. Citizen Villette, therefore, who still guarded the precious heart of his deceased friend, now set afoot a plan to rescue the corpse too from oblivion.

In the journal *La Chronique*, he first broached his scheme to translate Voltaire's body to Paris for honorable burial as a spiritual father of the Revolution. His suggestion caught on rapidly, and soon a public debate ensued concerning where the hero's body should be planted: beneath the altar of the Federation? Beneath the statue of Henri IV, whom he had immortalized in his poem *La Henriade*? At the end of the Champs-Elysées? In November 1790, Villette settled the matter himself. Suddenly ascending the stage during a performance of Voltaire's *Brutus*, he harangued the cheering audience and declared that Voltaire merited nothing less than to be interred in the new Panthéon, formerly the church of Sainte-Geneviève. To guarantee acceptance of this idea, he offered to pay all the expenses himself.

Paris, however, had competition. The commune of Romilly-sur-Seine, near the abbey, aware of the plight of Voltaire's bones, was making plans to rescue them. So was the Society of Friends of the Constitution, in Troyes. On May 10, 1791, representatives from both municipalities were present at Scellières for the exhumation, competing for their respective rights to the remains, when a decree from the Constituent Assembly arrived that settled the matter by declaring Voltaire to be the property of the state. Thwarted in their attempt to obtain the whole cadaver, the citizens of Troyes asked to be permitted to keep the skull as a relic, while the Romillians, representing an insignificant village, begged only for the right arm. Both requests were refused. However, in the confusion, someone stole the left heel bone.

The decayed corpse was stuck to the bottom of its rude coffin, so box and all were marched to the nearby village under guard to prevent any further desecrations. Nevertheless, another large part of the left foot and two teeth disappeared while the body was awaiting transfer to a more imposing coffin. The foot bones were on display for a time in the museum of Troyes, but they have long since disappeared. One tooth was given to Monsieur Charon of the delegation from Paris, and the other to a journalist named Lemaître to induce his silence about the theft of the foot. Lemaître used to wear the philosopher's tooth as an amulet; on his death it passed to his cousin, who was, appropriately enough, a dentist. No more is known of these teeth.

On May 30, the Constituent Assembly passed another decree authorizing the burial of Voltaire in the Panthéon. While plans were being laid in the capital for a grand funeral extravaganza, a huge and elaborate hearse was built, and at the beginning of July it set out for Romilly. The slow procession back to Paris occupied six days, stopping at every town and village along the way for speeches, parades, and salutes. Fresh wreaths and garlands were piled on the catafalque at every halt. On the evening of July 10 the cortege reached Paris and the coffin was set on a special platform erected on the ruins of the Tour de la Bazinère, the very tower of the Bastille in which the young François Marie Arouet, during his first imprisonment in 1717, had adopted the name Voltaire. Here it remained overnight, under a guard of honor.

The next day was completely taken up with a giant funeral procession, winding its way through the streets of the city to the cheers of hundreds of thousands of citizens. There were cavalry, infantry, delegates to the Assembly; workers, writers, children; statues, floats, and medallions; songs, banners, speeches, and slogans from one end of Paris to the other. It was a festive day, the people celebrating both the return of Voltaire in triumph and the return of their despised king in disgrace; for on the previous day, Louis XVI had been captured at Varennes, as he tried to flee the country, and had been escorted as a prisoner back to the capital.

The funeral car itself, thirty feet high with bronze wheels and drawn by twelve horses, stopped strategically at important sites, including Villette's house, the old Comédie-Française, and the Opera. When it reached the Odéon at seven in the evening, a heavy rain scattered the crowd, and Voltaire's bier was left alone in the downpour. When the storm abated, a small party escorted it to the Panthéon, where Voltaire was entombed in the crypt without further ceremony, not far from Descartes and Mirabeau.

Let us now turn our attention to Voltaire's brain. During all the years when Voltaire's body was in Scellières, his pickled brain floated placidly in Monsieur Mithouart's jam pot. After the apothecary's own death, his son and heir tried to give it to the Directory, then in control of affairs, so that it could be preserved as a national treasure. The Directory politely declined the gift. Mithouart *fils* put the jam pot away in a cupboard for a few decades. In 1830 he tried to give the brain to the restored royal government, again without success. Sometimes he would show his unwanted memento to visiting scientists, one of whom burned a piece of it in a candle flame just out of curiosity, but none were sufficiently interested to take it off his hands. His heirs tried to unload the horrid thing on the Académie Française in 1858, but the academicians begged off with the lame excuse that they had "no suitable reliquary in which to put this unexpected object." At the death of the last Mithouart about 1870, it was willed to a certain La Bosse, an employee of the family apothecary shop. He died in 1875, and his goods were auctioned. The records of sale omit to mention who bought the old jam pot and its contents, and nothing further is known about the fate of Voltaire's brain.

Voltaire's heart, however, is another story. The marquis de Villette held it in every honor. He sent it to Voltaire's estate at Ferney, where it was kept atop a fluted column in the philosopher's former bedroom. Later, Villette bought the estate and lived there for a while, but financial problems eventually led him to lease it to a rich Englishman. Having been criticized for renting Voltaire's heart along with the rest of the furniture, he transferred it to his ancestral estate at Villette, then to his house in the rue de Beaune in Paris, where it rested when Voltaire's funeral procession passed by in 1791.

Madame de Villette, who had adored Voltaire, preserved his honored heart after her husband's death. As the Revolution degenerated into chaos, she was forced to move several times, but eventually she returned to the Château de Villette with the heart. She turned the house into a sort of Voltaire museum and used the precarious immunity thus gained to provide a refuge for priests who were in danger from the excesses of the Reign of Terror. It was ironic indeed that the heart of Voltaire, who never was noted for his love of clergy, should have saved so many priests from a persecution that was inspired, in part, by Voltaire's own anticlerical writings.

Madame de Villette left Voltaire's heart to her son, who died childless in 1860 and left it to the bishop of Moulins, who, scandalized at possessing such a detestable relic, donated it to the nation. In 1864, the emperor Napoleon III decided that the most logical and decorous repository for Voltaire's heart was the same crypt in the Panthéon that contained Voltaire's skeleton.

Much to everyone's surprise, when Voltaire's coffin was opened, his skeleton was not there! A quick search of the Panthéon not only failed to find it, but Rousseau's turned up missing as well. It proved impossible to keep this embarrassing situation hushed up. A journalist exposed it and wrote that it was scandalous to let visitors to the Panthéon pay their respects at an empty tomb. The emperor ordered an investigation, and at last the police got the whole story from a man whose father had revealed it to him long ago.

It seems that fifty years earlier, in May 1814, when the monarchy had just been restored after Napoleon's exile to Elba, a group of reactionary young fanatics, led by the director of the royal mint, Monsieur de Puymorin, decided to cleanse the Panthéon of the irreligious remains of Voltaire and certain other undesirables that for too long had polluted the former church. Stealing into the temple by night, they opened Voltaire's coffin, threw his bones in a sack, and took them secretly to a dump outside the city. There, in a hole already prepared, they emptied their sack, pouring in quicklime to destroy all traces of the despised philosopher's body. They covered the hole with debris and dirt, leaving Voltaire where he had always feared to end up, cast away to rot on a rubbish pile.

There being no hope of recovering Voltaire's remains from beneath what had since become the site of the Halle aux Vins, Napoleon III decided to deposit the heart, all that re-

French officials place the rediscovered heart of Voltaire in the base of his statue in the Bibliothèque Nationale. (*Courtesy of Library of Congress*)

mained of the great Voltaire, in what he hoped was a safer place, the Bibliothèque Nationale. Alas, even the librarians, whose institution was so enriched with Voltaire's writings, proved remiss custodians of his heart, for they soon misplaced it. It did not surface until accidentally rediscovered in 1924, preserved in fluid inside a gilt box. This time it was placed within a special locked compartment in the base of the library's grinning statue of Voltaire, where, unless some new catastrophe has overtaken it, it still resides.

CHAPTER 23

Molière's Not There

Molière, the renowned French playwright and actor, has a small but impressive sarcophagus in the chic Père Lachaise cemetery in Paris. However, Molière himself is not there. The sarcophagus that bears his name does not contain his body, but rather the remains of some Parisian nobody. How comes it that the tomb of Molière is occupied by an impostor? And where is Molière, anyway?

Molière, the consummate actor, whose real name was Jean-Baptiste Poquelin, made a memorable final exit. On February 17, 1673, while playing the lead in a performance of one of his own plays, *The Imaginary Invalid*, he went into convulsions right on the stage. Since his role of Argan the hypochondriac required him to act the part of an ill man, he managed his sudden seizures so adroitly that the audience thought his coughing and trembling were part of the script. Later in the play, the plot required him to feign death; an hour after the performance, he experienced the real thing, suffocating from a severe hemorrhage of the lung.

Since he was an actor, the Church rules of the time automatically excluded him from burial in consecrated ground, just like heretics, sorcerers, usurers, and pagans. In his last moments, Molière sent for a priest in order to renounce his profession, make his confession, and receive the Last Sacrament,

This is not Molière's tomb, but everyone pretends it is. The one behind it is not really La Fontaine's either. (*Photo by Edwin Murphy*)

but it was too late. He died without formal repentance, and the vicar of the parish church of St. Eustache had no choice but to refuse him a Christian burial. His widow appealed to the archbishop of Paris for a special dispensation, but he was unhelpful. She then appealed directly to the king, who, although publicly doing nothing, privately sent word to the prelate "to avoid a disturbance and a scandal." By the time the archbishop relented, the corpse had lain scandalously unburied for four days. Even then, the archbishop imposed some restrictions:

> We authorize the vicar of St. Eustache to give ecclesiastical burial to the body of the deceased Molière in the cemetery of the parish on condition that there shall be no ceremony, with two priests only, after nightfall, and

that there shall be no solemn service for him either in the parish of St. Eustache or elsewhere, in any church of the regular clergy.

These conditions were mostly ignored. Molière was buried on the night of February 21, 1673, in the cemetery of St. Joseph in the parish of St. Eustache. Four priests carried the body, and three others officiated; children and lackeys, in livery, bore candles and torches, while a hundred of the deceased's friends attended with lighted tapers. A large crowd assembled for the event, and some sort of loud disturbance broke out, the exact nature of which is not known, though Molière's widow was obliged to scatter coins to the mob to restore tranquility. One contemporary source deliberately obscures the incidents surrounding Molière's interment, hinting darkly that there is much more to the story, but that he dare not print it because of "facts whose gravity imposes silence on everyone concerned." Rumors current at the time said that the actor was secretly removed from his consecrated grave, which the archbishop had permitted only to "avoid disturbance," and cast into the mass grave set aside for unbaptized persons.

At any rate, Molière finally was installed somewhere in the cemetery of St. Joseph, though whether in consecrated ground "at the foot of the cross" or in the squalid corner reserved for suicides, the stillborn, and the unbaptized is not clear. Even less clear is what happened after that. Numerous but conflicting services mention various movements of Molière's body. It seems probable that it was transferred into St. Joseph's church in 1750, but records differ about whether it remained there, was returned to its original plot, or reburied elsewhere in the cemetery. By 1792, when a certain part of Paris decided to rename itself La Section de Molière et de La Fontaine, and the revolutionary government decreed that as a reward, that section should have the bodies as well as the names of its new patrons, no one could tell exactly where in St. Joseph's cemetery Molière's remains could be found.

This technicality presented little or no problem to the two commissioners sent to claim the prizes awarded to their section. One decomposed skeleton being pretty much like all the rest, and there being no way to associate names with most of them, the commissioners simply dug up a likely candidate from the plot of the unbaptized and said it was Molière. Problem solved! They exhumed another body at random and called it Jean de La Fontaine, despite the fact that the poet had been buried in a different cemetery. The *faux* Molière, along with his companion "La Fontaine," for some reason were not reburied immediately, but were put in new coffins and stored for seven years, first in a crypt of St. Joseph's church, then in a guard house, until about 1799. Then they were removed to Alexandre Lenoir's museum of French monuments, located in the building that now houses the Ecole des Beaux Arts. Here their tombs were on display for eighteen years. By then, no one questioned their authenticity, and the supposed tomb of Molière was quite a popular attraction.

This popularity caught the attention of Nicolas Frochot, the promoter who practically invented the concept of the modern cemetery. When he opened his Père Lachaise cemetery on December 2, 1804, with its many innovations such as parklike, well-tended, and hygienic grounds, one person per grave, individual or family plots purchased in perpetuity, and individual grave monuments, it was the beginning of a new era in Western burial customs. However, like anything new, the garden cemetery concept was slow to catch on at first. Frochot spent 160,000 francs to acquire his seventeen-acre graveyard, and he sunk a lot more money into it to make it an agreeable place in which to be buried. Nevertheless, despite its aesthetic appeal, Parisians were not dying to get into Père Lachaise, at least not at a rate fast enough to suit Nicolas Frochot.

Racking his brains to find some way to attract corpses of the right sort (i.e., wealthy), he hit upon the idea that social climbing need not stop at death. Observing that people seek, during life, to be known by the company they keep, and are therefore at great pains to go to all the right places and be seen with all the right people, Frochot reasoned that he could capitalize upon this urge and, by a natural extension, convert it into a desire to be *buried* in the right place, among the right people.

To this end, he attempted to stock his mostly empty cemetery with as many celebrity corpses as he could conveniently acquire, as a subliminal advertising ploy. One of his first catches, in 1806, was the royal remains of Louise de Lorraine (died 1589), the queen of Henri III. Later, he traded Louise for the bodies of the celebrated medieval lovers Heloïse and Abélard, whom he buried romantically side by side. Eventually, he cast his eyes on the supposed cadaver of Molière, which he succeeded in obtaining from Lenoir's museum at the time of its suppression in 1817. La Fontaine's body was included in the bargain too. Frochot didn't know, and wouldn't have cared if he did know, that the bodies masquerading as Molière and La Fontaine were those of a couple of unidentified people, picked almost at random. The lucky substitutes were buried in style at Père Lachaise and are still there, enjoying the posthumous honors intended for their betters.

The real Molière's bones, meanwhile, probably remained somewhere in St. Joseph's cemetery. If not disposed of sooner to make room for fresh corpses needing a place to decompose, they were transferred to the catacombs of Paris, along with millions of other skeletons, when St. Joseph's and other old cemeteries were closed as a public health measure. The alleged tooth, jawbone, and vertebra of Molière, which at one time or another were honored as relics, cannot be authenticated with certainty, but probably came from the false Molière.

Frochot's gimmick paid off handsomely. More and more

rich bourgeois began to think of Père Lachaise, the resting place of so many notables, as the socially correct place to be buried. A man and his family could lie for eternity there, beneath the most ostentatious monument they could afford, side by side with the greats of history and the arts, and all it took was money. One's own tomb could be erected next to that of someone famous, like Molière, despite the fact that Molière is not really there.

The Mislaid Bones of John Paul Jones

J ohn Paul Jones is one of America's foremost naval heroes and is regarded as the father of the United States Navy. His exploits in the Revolutionary War are legendary, and his famous retort, "I have not yet begun to fight," is familiar to every American, exemplifying an iron determination that snatches victory from the jaws of defeat. Such widespread fame in our day, however, conceals the embarrassing fact that he died in obscurity in a foreign land and was so completely forgotten that when America belatedly wished to bestow on him a hero's funeral, his body could not be found. Its ultimate rediscovery was made possible only by the disinterested generosity and inspired forethought of an obscure French civil servant.

John Paul Jones was born in Scotland in 1747 and went to sea at the age of ten. His nautical career landed him in Virginia just before the American Revolution, in which he sided with the colonists. Congress commissioned him a naval officer in 1775, and awarded him a gold medal in 1787 for his victories over the British. Louis XVI, America's ally in the war, also made him a chevalier of France. Later, Catherine the Great of Russia hired him as rear admiral. After some victories against

John Paul Jones's body was lost for a century,
then had the most expensive funeral in
American history. (*Courtesy U.S. Naval Academy*)

the Turks, however, he fell victim to court intrigues and retired
on a leave of absence to Paris in 1790.

Paris at that time was swept up in the escalating French
Revolution. Jones offered his services as a naval commander,
but his career was in eclipse and his health was failing. At his

forty-fifth birthday on July 6, 1792, while awaiting a commission from the French government, he was already near death. Kidney disease and bronchial pneumonia had confined him to his lodgings when, on July 18, anxious friends insisted that he make his will without delay.

Since he was not a French citizen, his will would not be legal unless attested by Gouverneur Morris, the American minister to France. Morris brought notaries to whom Jones dictated his will. His estate consisted largely of uncollected debts and claims for compensation and prize money from the governments he had served. On paper, he was worth about $50,000. Unfortunately, he had little ready cash, and Gouverneur Morris, disregarding his past services to the cause of liberty, treated him like a washed-up indigent.

Morris scurried off to a dinner engagement, from which he was urgently recalled about 8:00 P.M. He arrived to find John Paul Jones already dead, and promptly directed Jones's landlord to dispose of the commodore's cadaver as cheaply and quietly as possible, an order for which he has been execrated ever since. There were, of course, some plausible excuses for his course of action. Since Jones was not Catholic, bureaucratic delays could be expected in obtaining permits for his interment in the only legal Protestant cemetery, four miles away outside the walls. Protestants who died in Paris were often buried surreptitiously to evade the law and avoid delay.

As to expense, Morris knew Jones had limited funds. As he later explained: "Some people here who like rare shows wished him to have a pompous funeral, and I was applied to on the subject; but . . . I had no right to spend on such follies either the money of his heirs or that of the United States. . . . I did not agree to waste money of which he had no great abundance, and for which his relatives entertain a tender regard." In fact, Morris auctioned most of Jones's uniforms, medals, and other paraphernalia to satisfy demands on his estate.

These considerations, however, do not exculpate Gouverneur Morris from the charge of callous indifference to the fate of John Paul Jones. Morris, as revealed in his letters and diary, had scant regard for Jones and treated the whole affair of his burial and estate as a nuisance. The minister of the United States government not only refused to honor the patriot's body, but declined to cooperate with those who would honor it. He was content that the hero's corpse should be hustled without ceremony into a pauper's grave.

The French had nobler ideas. Despite the turmoil of revolution and foreign war, the French government stepped in to rescue America's great naval warrior from an ignominious end.

Jones's friend Colonel Samuel Blackden on July 19 petitioned the Legislative Assembly in the name of Americans resident in France. "I announce to you that Admiral John Paul Jones died last night in Paris, and that the American Minister has ordered the person at whose house the Admiral lodged to cause him to be interred in the most private manner and at the least possible expense. This person, because of the formalities still existing concerning Protestants, found it necessary to apply to a commissary. . . . and M. Simmoneau the commissary expresses his astonishment at the Minister's order, and says that a man who has rendered such great services to France and America ought to have a public funeral. He adds that, if America will not pay the expense, he will pay it himself. . . ."

Pierre-François Simmoneau, a royal commissary of Louis XVI with jurisdiction over the Protestant cemetery, did more than pay the bill and cut through red tape. Far more important, he provided for the day when America, repudiating the scandalous indifference of its minister, should wish to restore the patriot's corpse to its rightful honors. He preserved the body with alcohol in a lead coffin so that "in case the United States should claim his remains, they might more easily be removed." This precaution, the clue to which was found in one of Blackden's letters, was crucial to the rediscovery of the body 113 years later.

The Assembly also voted a public funeral for the following evening, featuring a procession from Jones's lodgings to the St. Louis Protestant cemetery in the suburbs. The cortege was accompanied by a detachment of grenadiers and a twelve-man deputation from the Assembly.

With muffled drums, a volley of musketry, and a stirring oration, John Paul Jones was lowered into his grave. Some friends and neighbors, five Americans, and a few old shipmates constituted the only audience besides the official delegation. Gouverneur Morris was formally invited but, with gauche insensitivity, he declined on the excuse that "he had persons to dine with that day." His diary reveals that after dinner, while the funeral was yet in progress, Morris frittered away the evening making vapid social calls on the wives of prominent officials.

Although the generous M. Simmoneau probably expected America to reclaim the body within a year or two, circumstances militated against it. Even before news of Jones's death reached America, a mob stormed the Tuileries palace. Swiss Guards massacred in defense of the king were dumped into mass graves all around Jones's plot in the Protestant cemetery. The Allied invasion of France resumed in earnest on August 19, fighting intensified on all fronts, and the political situation deteriorated into the Reign of Terror. If anyone in America entertained the hope of repossessing John Paul Jones from foreign soil, the chaotic international situation would scarcely have permitted it. No one in the United States, however, was unduly concerned about Jones or where he was buried. Although remembered by historians, John Paul Jones was all but forgotten by the general public as the years rolled by. Congress tried desperately to forget him, but importunate claims by his heirs for unpaid salary and prize money wouldn't permit it to do so. In lieu of money, Congress authorized naming a ship after him in 1834, but with the government's customary dilatoriness the ship was not actually commissioned until 1862. Better late than never, Congress finally made a partial payment to the heirs in 1848, and paid the remainder with interest in 1861.

The first serious proposal to fetch Jones's corpse came from secretary of the navy George Bancroft in 1845. By then, however, no one remembered exactly where to find it, so nothing came of the idea until John Henry Sherburne, register of the Navy Department, revived the project in 1851. Sherburne, the son-in-law of one of Jones's officers, also gave up the attempt without much progress. The quest languished until 1899, when the journalist Julius Chambers gave it wide publicity. He claimed (erroneously) to have discovered the gravesite, and he prepared a resolution for Congress on the subject. Congress ignored the resolution, but the interest raised by Chambers encouraged Horace Porter, American ambassador to France, to persevere in the search at his own expense.

His first problem was to identify the right cemetery. In the intervening century, discrepant traditions had arisen that John Paul Jones was interred in Père Lachaise cemetery or in St. Martin's, the Protestant burial ground that preceded St. Louis. When old records revealed that St. Louis was the correct cemetery, there was the further obstacle of determining its location. It had closed in 1804, and the expanding city obliterated all trace of it. Old maps finally pinpointed the site, but the ground was found to be covered with buildings, whose owners demanded exorbitant sums for excavations on their property.

Porter postponed operations until the excitement died down. In 1905, taking advantage of a slum clearance project, he at last secured the necessary permissions and commenced underground explorations. Gangs of workers toiled around the clock making soundings, sinking shafts, and digging tunnels. The unfortunate Swiss Guards, badly decomposed, were found "stacked like cordwood." After three weeks, the position of the grave had been ascertained to within a few feet. All that remained was to find a lead coffin.

That was no problem. Five lead coffins surfaced within those few cubic meters. By this time Porter must have been tearing his hair. Fortunately, three of the boxes were neatly labeled with the names of their occupants. But the other two were not, so a dozen anthropologists and pathologists were called in. After six days, relying on busts, portraits, and descriptions of John Paul Jones and accounts of his final illness, the scientists definitely eliminated one body and made almost positive identification of the other. The alcohol in which it had been buried, besides preserving it in remarkable condition, was also a decisive clue to its identity. Simmoneau's inspired precautions had at last borne fruit.

Porter's find was a sensation. He had redeemed the honor of the United States and remedied the dereliction of his predecessor, Gouverneur Morris. President Roosevelt was notified immediately. Sensing the publicity value of the find in his drive to make the United States a great naval power, he ostentatiously dispatched an entire battle fleet across the ocean to bear the precious cargo home, thereby setting in motion the most expensive series of funeral celebrations in American history. The French, not to be outdone in enthusiasm or display, quickly arranged elaborate public obsequies on July 6, the anniversary of Jones's birth. Then the old lead coffin, newly encased in a sheath of mahogany, was escorted by military parade to the depot, whisked by special train to Cherbourg, and piped aboard a French torpedo boat for transfer to the USS *Brooklyn*. A French naval squadron accompanied the American fleet back across the Atlantic.

Nine American battleships joined the armada off Nantucket and followed John Paul Jones to Chesapeake Bay, where thunderous fifteen-gun salutes signaled his approach to the Naval Academy at Annapolis, Maryland. He arrived on July 22, 1905. By this time patriotic hoopla was rising fast, as every city remotely connected with the once-forgotten John Paul Jones now vied for possession of his pickled corpse. But Roosevelt insisted that the Naval Academy should be its final resting place. With another reverent ceremony, the casket was deposited temporarily in a special vault until a really splendid funeral could be arranged, one at least as grandiose as the sendoff provided by the French.

This took time, but the big event finally occurred on April 24, 1906, attended by dignitaries from around the world. The festivities went on all day, the big-navy program got the boost that Roosevelt desired, and everyone went home smiling. Then the guest of honor was ignominiously stowed out of sight behind the stairs of Bancroft Hall, and there he remained for seven more years!

Congress, it seems, having agreed that John Paul Jones should be enshrined forever in a lavish crypt beneath the Na-

The sarcophagus of John Paul Jones at the U.S. Naval Academy. (*Courtesy U.S. Naval Academy*)

val Academy chapel, then penuriously declined to appropriate any money to build it. Public indignation, fanned by the press, finally forced it to act. At long last, on January 26, 1913, John Paul Jones underwent his fifth and final funeral. He was laid to rest in a magnificent porphyry sarcophagus within a marble crypt modeled loosely on Napoleon's tomb.

America's long-lost admiral is now enshrined in the glory he deserves, while Gouverneur Morris is buried in equally fitting obscurity at a place known only to antiquarians.

121

All Paine, No Gain: Tom Paine's Stolen Remains

Thomas Paine was one of the most influential radicals of the modern era. For his passionate, polemical, uncompromising championship of the dignity of men against the tyranny of governments, the entire world owes him a debt of gratitude. In hindsight, his works such as *Common Sense*, *The Rights of Man*, and *The Crisis* are recognized as classics in the development of the modern concept of human rights. Yet he was such an inveterate controversialist, had such a knack for offending people and getting into trouble, and held such unpopular religious beliefs, that by the time of his death in 1809, he had few admirers, fewer friends, and legions of enemies in England, Europe, and America. As death approached, he expressed fear that his adversaries would never let his body rest in peace.

He was wrong. His enemies left his body alone. Instead, it was one of his most ardent followers who robbed his grave, hauled his corpse all the way across the Atlantic, submitted it to a ridiculous public exhibition, and then left it to his bankrupt heir, who eventually lost it. With friends like this, who needs enemies? The tragicomic story of Tom Paine's remains

Thomas Paine. His body was stolen by William Cobbett and taken to England for a crazy fund-raising scheme. (*Courtesy of Library of Congress*)

is yet one more object lesson in our study of what can happen to someone *after the funeral*.

A little background about Tom Paine is in order. Despite being one of the founding fathers of the United States of America, Tom Paine was not American by birth or upbringing. Like John Paul Jones, he never set foot in North America until just before the Revolution. Born in England in 1737, he was largely self-educated. Benjamin Franklin, during one of his visits to England, took an interest in young Paine and encouraged him to emigrate to the colonies. His embarrassed financial situation also induced him to abandon his government excise post and, one step ahead of his creditors, take ship for America in 1774.

He quickly became editor of the *Pennsylvania Magazine*. On January 10, 1776, he published his famous tract *Common Sense*. This remarkable pamphlet created an immense sensation and was largely responsible for changing a rebellion of colonies into a war for independence. Paine held various military, state, and national posts during the Revolutionary War, for which Congress rewarded him with gifts of money, and the state of New York gave him a large farm at New Rochelle.

Tom Paine returned to England in 1787 to promote republican and popular causes. His celebrated defense of the French Revolution, *The Rights of Man*, came out in 1791. This earned him both an indictment for treason in merry olde England and a seat in the French convention; for, even though he was a foreigner and could not speak French, he was elected to represent the department of Calais on the strength of his radical opinions.

His popularity with the Jacobins didn't last long, and Robespierre had him imprisoned and almost guillotined. He whiled away his incarceration by writing tracts that offended just about everyone in Europe he hadn't succeeded in offending before. Later he vilified George Washington's policies and military reputation, out of resentment that America had not done enough to save his neck. Thus, when he finally returned to New York in 1802, he was distinctly persona non grata. At his death in 1809, having been denied a place in the Quaker cemetery of New York, he was buried at a spot of his own choosing on his farm. Only six people were present at the interment twenty-two miles from New York City. In accordance with his will, a wall twelve feet square was raised around the

Thomas Paine's farm in New Rochelle, New York. Paine's body was buried here before Cobbett stole it. (*Courtesy of Library of Congress*)

grave, which was flanked by two weeping willows and two cypresses. A sober grave marker, with his age misstated, identified him as the author of *Common Sense*. Only one newspaper, the *New York Post*, mentioned his demise, noting wryly that "he had lived long, done some good, and much harm."

Despite minor desecration of the gravestone and the trees, attributed to souvenir hunters rather than enemies, nothing disturbed Tom Paine's rest until the advent of "Peter Porcupine" ten years later.

"Peter Porcupine" was the pen name of William Cobbett, the English pamphleteer and radical reformer who almost singlehandedly invented popular journalism. Originally a staunch conservative, during his previous sojourn in America from 1792 to 1800 he revelled in publishing virulent tirades against anyone more liberal, progressive, democratic, or radical than he was—which included nearly everyone. Tom Paine was one of his special targets, and his scurrilous *Life of Tom Paine* was instrumental in blackening Paine's reputation with the few people that Paine himself had not already alienated.

Cobbett was one of the most interesting men of his era. As full of energy as an exploding bomb, and equally ungovernable and unpredictable, he has been characterized as "a boisterous Englishman, with an acute, untrammeled, robust, and untrained mind, which he changed over night, with apol-

William Cobbett. After failing in his plan to raise money by exhibiting Paine's body, he left the cadaver to his bankrupt son, who lost it. (*Courtesy of Library of Congress*)

ogies to none. A man of extraordinary virility, of monumental nerve and impudence, he excelled his whole generation in ferocity of invective. . . . There was a Napoleonic quality in the man's insolence and courage."

Cobbett's mercurial and pugnacious personality was consistent throughout his long career, but his political views underwent a surprising and complete reversal after his return to England. Shocked by the impoverishment of the rural labor class brought on by the Industrial Revolution, Cobbett quickly shed his conservative principles and became one of the most extreme radical reformers in the United Kingdom. It was in this frame of mind that he reread Thomas Paine's *Decline and Fall of the English System of Finance*, which he had formerly excoriated in his vituperative writings. Now, however, the scales having dropped from his eyes, he viewed Paine's work as profound, and Paine himself as a prophet.

If Paine was the prophet, Cobbett would be his apostle. He announced to the world: "Old age having laid its hand on this truly great man, this truly philosophical politician, at his expiring flambeau I lighted my taper." Cobbett found that being a radical reformer in a monarchy ruled by the ultraconservative propertied classes was not as congenial as being a defender of the establishment. Never one to do things half-heartedly, he preached his new views so energetically that he soon made himself obnoxious to the power structure whose darling he had once been. He spent two years in prison for sedition, and finally fled to America in 1817 to avoid further prosecution.

Here, among many other frenetic activities, he persisted in his plan to rehabilitate Tom Paine's reputation and vindicate his views. It wasn't easy. In England, the conservative press had so successfully vilified Tom Paine that he was almost as unpopular as Napoleon, and many towns burned him in effigy on Guy Fawkes Day in preference to Guy Fawkes. In the United States, the children sang this ditty:

Poor Tom Paine! Here he lies,
Nobody laughs and nobody cries;
Where he's gone and where he fares,
Nobody knows and nobody cares.

William Cobbett wrote another biography of Tom Paine, this one full of praise, but never got around to publishing it. Instead, he was diverted by an even more ingenious idea, at least in his opinion, for reviving the fame of his hero. At the same time, it would rescue Paine's body from the shameful neglect into which ungrateful Americans had allowed it to fall. As Cobbett wrote:

Paine lies in a little hole under the grass and weeds of an obscure farm in America. There, however, *he shall not lie, unnoticed, much longer.* He belongs to England. His fame is the property of England; and if no other people will show that they value that fame, the people of England will.

America was not worthy to be the final resting place of the great Englishman Thomas Paine, and the impetuous Cobbett was just the man to do something about it. He would take Tom's body back home to England and rebury it in style, with a spectacular funeral and a wonderful monument. The grave would become a rallying point for the poor and downtrodden masses.

Cobbett lacked only two things requisite for carrying out his plan: permission to take the body and money to build the monument. As to permission, he simply dispensed with it. By visiting New Rochelle frequently and standing with bowed head beside Paine's grave, he accustomed the local populace to his presence on the farm. So, when he was seen in the neighborhood in late September 1819, with a wagon and two other men, no one was suspicious until they noticed, the next day, a large hole in the ground where Tom Paine's coffin used to be. The sheriff of New Rochelle somewhat tardily set out in pursuit of the grave robbers, but lost the trail at Yonkers. Cobbett reached New York and booked passage for England, his contraband cadaver concealed in the hold of the ship as a crate of merchandise.

The other thing Cobbett lacked to accomplish his projected mausoleum for Paine was money. His fortune and credit depleted by years of prison and exile, he needed a new source of funds to bury Paine with honor and finance his liberal program of reform. Ever resourceful, he hit upon what he was sure was the right plan. He would take the corpse on a fund-raising tour which would not only pay for a grandiose tomb and a splashy funeral with "twenty wagon loads of flowers . . . to strew before the hearse," but would also serve as a revolutionary catalyst for reform. Cobbett's fertile mind fairly ran riot with imaginary possibilities. As he himself expressed them in a letter to J. W. Francis:

I have done myself the honor to disinter his bones. I have removed them from New Rochelle . . . they are now on their way to England. When I myself return, I shall cause them to speak the common sense of the great man; I shall gather together the people of Liverpool and Manchester in one assembly with those of London, and those bones will effect the reformation of England in Church and State.

This prediction was wildly optimistic, to say the least. On first arriving at Liverpool, the defunct Thomas Paine created quite a stir. According to newspaper reports, a large crowd met Cobbett at the docks and "cheered him to the inn." Later at the customhouse yard Cobbett opened the box of relics to give the crowd a peek. "There, gentlemen," he solemnly intoned, "are the mortal remains of the immortal Thomas Paine." Instead of an outpouring of spontaneous popular enthusiasm, however, Cobbett was surprised to encounter nearly universal scorn and derision for his pet project. Only the government failed to greet his plan with the laughter it deserved. Instead of scoffing like everyone else, the government, unamused, persecuted and obstructed. Cobbett was denounced in Parliament. Even a town crier, who spread the news of the bones' arrival in Liverpool, was jailed for nine weeks. The government naturally refused permission for the weird scheme of exhibiting the skeleton around the country to raise money.

Cobbett stubbornly persisted. He held "bone rallies" in the Midlands and London anyway. They were poorly attended. His newspaper, the *Weekly Register*, appealed for a public subscription to build Paine's monument. The response was underwhelming. He exhibited the corpse. Almost no one came. He scheduled a public dinner to celebrate Paine's birthday on January 29, 1820. The dinner was canceled. Cobbett was reduced to selling locks of Tom Paine's hair to raise a few shillings. The harder he tried to interest people in Tom Paine, the more he was the object of ridicule. The government-controlled press heaped him with abuse, and cartoons in the magazines lampooned him without mercy. Even poets took potshots at him. Lord Byron penned:

> *In digging up your bones, Tom Paine,*
> *Will Cobbett has done well;*
> *You visit him on earth again,*
> *He'll visit you in hell.*

Finally even the indefatigable Cobbett was forced to admit his grandiose ideas about Tom Paine were not widely shared. He shelved the plans for a monument, gave up on the funeral, and regretfully shoved the box of bones under his bed. In 1833 he sent them to his farm in Surrey.

There they stayed until his death in 1835. Bankrupt, he left the bones to his son, who was subsequently arrested for debt. The receiver insisted on auctioning them with the rest of the estate, but the auctioneer balked out of scruple. The ridiculous situation was resolved only when the son appealed to the lord chancellor, who ruled that the skeleton of Thomas Paine was not a marketable asset. The final insult!

What happened to Paine's skeleton after that has not been settled with certainty. Apparently it remained on Cobbett's farm until at least 1844. It has variously been said to have been buried in 1849 in the nearby churchyard of Ash or in the garden of a descendant of Cobbett. There are unsubstantiated allegations that it was owned at various times by Cobbett's secretary Benjamin Tilley (who extracted and kept part of the brain), by a day laborer named George West, and by a furniture dealer in London. The skull was rumored to have been detached and come into the hands of either a Brighton phrenologist or an Anglican clergyman. In 1854, a Unitarian clergyman, the Rev. Robert Ainslie, claimed to own Paine's skull and right hand, which were examined by Professor John Marshall of the Royal College of Surgeons (he found them "small and delicate" for a man). After that, the skeleton and its fragments vanish from the historical record. All that verifiably remains of the most widely read author of his time is some locks of hair that the self-proclaimed apostle, William Cobbett, sold from the head of his deceased prophet in a last desperate attempt to raise money for his tomb.

Daniel Boone's Bones

D aniel Boone just couldn't stay in one place too long. He was the quintessential American frontiersman, always moving, always seeking a new place for a farm, a new hunting ground, a new start. Boone was one of the restless sort, and he needed elbow room. Whenever he could see the smoke from another settler's cabin, he reckoned the neighborhood was filling up, and it was time for him to move on farther west. He made it as far as Missouri before old age finally obliged him to settle down. There, near Charette, he died in September 1820. But *after the funeral*, Daniel Boone made one last move—in the wrong direction!

His goal in life had been to keep moving forward, but after he died, some enterprising citizens of Kentucky had other ideas. In 1845 they bamboozled Boone's descendants and tricked the state of Missouri into letting them take his remains back to Kentucky for honorific burial. Besides being a move backward, it was a move of which Daniel Boone himself would scarcely have approved. For, despite being one of the pioneering founders of Kentucky, Boone felt that he had been treated badly there, and in later life he had often expressed the determination never to set foot in that state again, much less be buried there.

To understand why Daniel Boone's body was practically

Daniel Boone's body was taken from Missouri to Kentucky against his wishes. (*Courtesy of Library of Congress*)

hijacked from its grave on Tuque Creek, in Missouri, and hustled back to Kentucky, a quarter of a century after he died, you have to know something about the economics of cemeteries. Although cemeteries are familiar to modern civilization, they are a relatively recent invention. It was partly because of overcrowding and deplorable hygienic conditions that the traditional burial grounds in churchyards and family gardens were

133

Boone's original burial site on Tuque Creek, near Charette, Missouri. (*Missouri Historical Society, St. Louis*)

gradually replaced in the nineteenth century by the spacious, well-tended, landscaped cemeteries that twentieth-century city dwellers take for granted. The "cemetery movement" also was accelerated by changing attitudes about how the dead ought to be treated, with concepts like embalming, "perpetual care," scenic venues, individual plots, and imposing head-stones gradually coming to the fore.

The cemetery movement, once it got started, also was pro-pelled by civic pride and speculation. Starting with Père La-chaise cemetery in Paris in 1804, the practice of beautifully planned upscale cemeteries soon spread. London's Kensal Green cemetery opened in 1832, and Highgate in 1839. The fad reached America in the same decade, which saw the estab-lishment of Mount Auburn cemetery in Boston, Laurel Hill in Philadelphia, and Greenwood in Brooklyn. Whatever was in vogue on the eastern seaboard was sure to be emulated by the burgeoning cities and towns of the interior. Imitation led to rivalry, and cities vied to have the biggest, best, and prettiest cemetery they could manage. Medieval towns had competed in the magnificence and size of their cathedrals, and fin de siècle metropolises later gloried in opera houses and train stations, while modern megalopolises splurge on imposing airports and sports arenas. In the mid-1800s, however, the civic rage was for cemeteries.

Frankfort, the capital of Kentucky, was not immune to the mania for planned garden cemeteries. In 1845, some entrepreneurs in that fair city had a problem on their hands. Having incorporated themselves as the Capital Cemetery Company, they had acquired a perfect tract of land on a scenic hill overlooking the town and were busy improving their site to attract business. But scenic views, landscaping, and pleasant walkways weren't necessarily enough. To make this a cemetery that people were dying to be buried in, the promoters knew that they must also appeal to the basic human instinct that the sociologist Thorstein Veblen later called "invidious comparison," or, in plain English, status-seeking. People who derive a vicarious status by rubbing elbows with the rich and famous during life should be allowed to continue this pastime after death by being buried near the rich and famous. Imitating the successful marketing technique used by the promoters of Père Lachaise and similar big-city celebrity cemeteries, the sagacious owners of the new Frankfort cemetery were on the lookout for prestigious corpses that they might acquire and use as social magnets to attract the fee-paying bodies of their status-seeking fellow citizens.

Since cemetery impresarios all over the country were doing the same thing, most of the best available cadavers, such as Revolutionary War heroes and founding fathers, had already been claimed. Then someone remembered Daniel Boone. A little discreet investigation revealed that the great pioneer of the Cumberland Gap, the founder of Boonesboro and several other Boone-towns in Kentucky, was buried in disreputable obscurity on a backwoods farm in Missouri. Daniel

Boone would be an ideal candidate for celebrity burial in the capital of Kentucky.

The cemetery promoters quickly began lobbying the state legislature, which in April 1845 obligingly passed a resolution authorizing the burial of Daniel Boone in the new capital cemetery. Of course, the justification for moving Boone's body was not expressed in terms of improving the financial outlook of the cemetery project. No, it was a matter of pure patriotic pride. Since Missouri had insulted the memory of the great Kentuckian Daniel Boone by failing to give his body the honorable burial it deserved, then Kentucky must act. If, twenty-five years after Boone's death, Missouri had not so much as erected a simple tombstone, it had forfeited its right of custodianship.

State leaders in Kentucky, at the instigation of the cemetery owners, began a letter-writing campaign to their Missouri counterparts and to Boone's descendants to convince them that Daniel's remains would be better off in Kentucky. These overtures worried the Missourians, whose officials quickly appropriated $500 to raise a monument over Boone's grave.

The Kentuckians had to act quickly, before opposition had time to consolidate. They wrote to Nathan Boone, Daniel's son, asking for his consent to the removal of his father's bones and promising to spend the incredible sum of $10,000 to make a suitable home for them in Frankfort. Nathan was absent on a military campaign, so his consent apparently was not obtained. In any case, the Boone clan no longer owned the farm where their patriarch was buried. The promoters therefore enlisted the aid of Daniel Boone's elderly nephew, William Linville Boone of Kentucky, and sent him with fellow citizens Jacob Swigert and Thomas Crittenden, minor Kentucky politicians, to deal in person with whomever they could and bring Daniel Boone's corpse back with them as soon as possible.

136

Apparently having first obtained the approval of two of Daniel Boone's granddaughters, both of whom were married to prominent Missourians, the Kentucky trio appeared unexpectedly on July 17, 1845, at the home of Harvey Griswold, who owned the so-called Bryan farm on Tuque Creek, where the remains lay. Griswold resisted their request to exhume the bodies of Daniel Boone and his wife, Rebecca. He first alleged that the deceased should not be removed from the place of rest Daniel himself had selected. He also objected to "any act which may deprive Missouri of the credit of doing appropriate honors." Besides, Griswold had paid "an extravagant price" for the farm because it was Boone's resting place, and removal of the bones would lessen its value. The promoters countered these arguments by presenting the written statements of Boone's granddaughters, by promising to compensate him for any loss of property value, and by also agreeing to erect a memorial monument on the Bryan farm commemorating Boone's original grave (a promise never kept).

The negotiations with Harvey Griswold dragged on all morning, and the news started to filter out. A crowd of neighbors drifted over to the farm, including some Boone relatives who had not been consulted, so the hesitant Griswold stipulated that their consent was essential before he would permit disinterment. Crittenden addressed the three dozen people who were milling around. What happened next has been the subject of much debate. Some of the bystanders objected to removal of the corpses, others didn't care, and some assisted the Kentuckians in identifying where to dig among the overgrown and poorly marked burial plots. In any case, no one actively interfered in the excavations, which were quickly done.

Almost nothing was left in the rotted coffins (Boone, in anticipation of death, had made his own coffin out of hardwood and kept it around the house until he needed it). The flesh was gone, and many of the bones disintegrated when moved. What survived was taken away in wooden boxes, except for stray teeth and bone fragments that some of the locals grabbed as souvenirs. The Kentuckians bade a hasty farewell and got away so fast that they did not even take time to replace the earth in the empty graves. Their unseemly haste, their dubious manner of securing the family approvals, and their unfulfilled promises to the Boones and Griswolds led to later allegations that Kentucky filibusters had all but stolen the remains of Daniel Boone and Rebecca from Missouri.

Back in Kentucky, at a ceremony in the capitol on September 13, prominent citizens watched the transfer of the bones to new coffins. They passed around Daniel's skull to admire it, and made a plaster cast of it so the local phrenologists (they were everywhere in those days) could study it at their leisure. The next day, four white horses drew a hearse bearing the coffins along a gay parade route from the capitol to the new cemetery. Bands played, the militia marched, thousands cheered, and orators orated. The Boones were buried in state. The following month, having gotten all the attention they could have desired, the cemetery moguls started selling lots. Everybody was happy.

Well, not quite everybody. As months and years passed without the cemetery owners fulfilling the promises of their agents to place a memorial marker on Bryan's farm or construct a monument over Boone's bones in Frankfort, the Missourians became increasingly resentful. They felt they had been duped by Kentucky swindlers who, having made them feel ashamed to have neglected the old grave of the revered Daniel Boone, were now neglecting the new one themselves. Even when the embarrassed Kentucky state government stepped in fifteen years later and put up a handsome monolithic cenotaph with bas-reliefs, the Missouri critics were not stilled: souvenir hunters had so defaced the monument that the names were barely legible a few years later. Even worse, the

Boone's grave marker in the cemetery overlooking Frankfort, Kentucky. (*Caufield & Shook Collection, Photographic Archives, University of Louisville*)

workmen employed in erecting the monument had carelessly mishandled the remaining bones, allowing at least one by-stander to swipe a vertebra for a personal relic.

Indignant Missourians started rumors that the bones removed from Tuque Creek in 1845 were not Daniel Boone's; they claimed the Boones and Bryans, to get rid of the pesky Kentuckians, had shown them the wrong grave. This tale is not supported by contemporary accounts, but it was widely believed; the Missouri legislature even varied its routine of periodically demanding the return of Boone's body, and began to demand only the return of Rebecca's, to be reburied beside her husband in Missouri. As late as 1983 the controversy flared anew when the Kentucky state forensic anthropologist, having studied the 1845 plaster cast of Boone's purported skull, proclaimed that it did not resemble a Caucasian cranium at all;

he said it could very well have been that of one of the black slaves interred in the Boone-Bryan burial ground.

The dispute continues to simmer. Now it is frankly admitted to be a matter of competition for the tourist trade as well as one of state pride. In 1987, for instance, the Warren County government, where Tuque Creek is located, petitioned the governor of Missouri to proclaim publicly that Daniel Boone's bones had never been removed from the state. Tourism officials had noticed a remarkable increase in tourism at *both* sites just after the widely publicized findings of the Kentucky state anthropologist a few years earlier. Like Christopher Columbus, who is buried in four different places, Daniel Boone has become a valuable tourist attraction; too valuable, in fact, to be buried in just one place.

Buffalo Bill's Burial:
Pike's Peak or Bust

D aniel Boone was not the only famous American whose resting place was disputed between different states. The bodies of two well-known figures of the last days of the Old West, Buffalo Bill Cody and the Sioux chief Sitting Bull, each became mired in controversy *after the funeral*. These two colorful characters, who were opponents during the Indian wars, later became colleagues, and cemented their friendship on the entertainment circuit; much later, both were cemented into their respective graves to settle decades of arguments over where they should be buried. Both stories are interesting. Let's start with the posthumous adventures of Buffalo Bill.

William Frederick Cody was his real name, but early in his career as a buffalo hunter on the Great Plains he acquired the nickname that stuck with him forever after. Besides killing over seven thousand buffalos, he was also an energetic and courageous cavalry scout and Indian fighter. His greatest talent, however, was as a clever showman and tireless self-promoter, the Phineas T. Barnum of the buckskin set. This is why Buffalo Bill is still remembered by the public at large, while

Sitting Bull and Buffalo Bill cementing their friendship. Later, each was cemented into his grave to discourage grave robbers. (*The Artist's Proof, Inc.*)

scores of equally capable buffalo hunters, cavalry scouts, and Indian fighters are long forgotten.

Buffalo Bill hit the big time with his fabulous Wild West show, which he started in 1883, an entertainment phenomenon the likes of which the world had never seen. At its peak, it included hundreds of Indians, cowboys, and other performers, as many horses, and fifty live buffalos. It thrilled audiences all over America and Europe, including monarchs and princes; even the staid Queen Victoria of England, a virtual recluse for twenty-five years after her husband's death, was exhilarated by Buffalo Bill's incomparable show.

Before he called it quits in November 1916, Bill had made millions. However, he had lost all those millions, and more, through bad investments. By January 1917, he was in financial distress despite his enormous popularity, and he was near death from uremic poisoning. His doctor suggested that he try the therapeutic waters of Glenwood Springs, Colorado, so he went there with his wife, Julia. It didn't help. His doctor then took him to his sister's house in nearby Denver, where he died on January 10. His last days were covered by newspapers all over the country, and his death caused nationwide grief, for Buffalo Bill was a much beloved hero in those far-off days when hero worship was not yet scoffed at. Telegrams and letters of condolence poured in from around the country and the world, including tributes from the king of England, the Kaiser, and President Wilson.

Bill unloaded two surprises on his friends and relatives just before he died. At his unexpected request, he was received into the Roman Catholic Church on his deathbed, despite the fact that none of his relatives were of that persuasion. His other big surprise was to ask, almost with his last breath, that he be buried on Lookout Mountain, the imposing Rocky Mountain peak looming just west of Denver. This request was quite strange, since Cody had long ago picked out his own grave on Cedar Mountain, near the town of Cody, Wyoming, which he had founded. His wish to be buried on Cedar Mountain was well known to all, and even was included in his will:

> It is my wish and I hereby direct that my body shall be buried in some suitable plot of ground on Cedar Mountain. . . .
>
> I further direct that there shall be erected over my grave, to mark the spot . . . a monument wrought from native red stone in the form of a mammoth buffalo, and placed in such a position as to be visible from the town. . . .

Julia Cody wrote years later that Bill simply changed his mind at the last second, and that she honored his request. This explanation seemed implausible to most people. Rumors, never proved, persisted that Harry Tannen, a Denver business and civic leader, was the instigator of the idea after Bill died. Sensing the tourist bonanza that, with proper management, might accrue to the Denver economy if the renowned Buffalo Bill were buried there, he allegedly offered the financially embarrassed and grief-stricken widow $10,000 for her acquiescence in the idea. Certainly the city fathers of Denver leapt at the chance to play host to Buffalo Bill. The mayor and the park commissioner immediately donated the most scenic spot they could find on Lookout Mountain. The Colorado legislature quickly authorized Buffalo Bill's body to lie in state in the capitol rotunda, for which the local U.S. army post and the Colorado National Guard provided a guard of honor. The Elks, the Masons, and other fraternal organizations cooperated in a splendid funeral procession, featuring the governors of two states, thousands of marchers, seventy old cowboys, and the dead hero's riderless white horse, stirrups reversed and pistols hung on the saddle horn. After the funeral Mass, said by the priest who had baptized Cody six days earlier, the embalmed body was temporarily put in cold storage. The ground on Lookout Mountain would be frozen until summer; not before then could a suitable burial be attempted, especially on the gigantic scale being planned.

The big event took place on June 3, 1917, when twenty-five thousand people accompanied Buffalo Bill's coffin on its trip to its panoramic grave site high on Lookout Mountain. The weary crowd dwindled to ten thousand before the summit was reached. There the glass-topped coffin was opened one last time for viewing, and after suitable speeches and ceremonies it was lowered into a shaft which had been blasted out of the solid rock.

Within a few years, a Buffalo Bill museum and wild west center had been built around the grave and its native stone marker. The tomb was as popular with sightseers as Washington's in Virginia and Grant's in New York, despite its remote location and the arduous ascent to see it. Those who had fore-

Buffalo Bill's coffin just before his interment on Lookout Mountain. (*Denver Public Library, Western History Department*)

seen the tourist potential in Buffalo Bill's bones were not disappointed.

Those who *were* disappointed were the good citizens of North Platte, Nebraska, where Cody had lived, and Cody, Wyoming, where he was supposed to have been buried. They suspected that sinister people in Denver had taken advantage of the accident that Bill had died in their town to hijack the body and bury it on their mountain, just for the tourist trade they hoped to engross by exploiting their windfall to the hilt. Had the exploiting not been such an obvious success, the other towns might not have cared so much. But as things stood, they felt that somehow they had been cheated out of their natural rights to exploit the body themselves. Both of the aggrieved towns built memorials to Buffalo Bill, but neither attracted much outside attention or tourist money.

The resentment of the two towns smoldered for a long time, and there were dark whispers and shadowy suspicions that something was going to be done to retrieve Cody's body for one or the other of Denver's jealous rivals. Just to make sure that potential grave robbers would be thwarted, Denver weighted down the buried bronze casket with several thousand pounds of concrete, reinforced with iron. A Denver newspaper defied Wyoming and Nebraska with a cartoon captioned "They shall not remove him from the resting place he chose." As if to emphasize that argument, Buffalo Bill's gravestone on Lookout Mountain includes the touching sentiment *At Rest Here By Request*, although skeptics and cynics note that the inscription cleverly avoids specifying whose request it was.

Even as late as 1948 the feud continued. In that year, the American Legion Post of Cody offered a $10,000 reward for the "return of Buffalo Bill's body from Colorado to Cody, Wyoming." In response, the American Legion Post in Denver mounted a guard over the grave until the city of Denver unburied the casket, blasted its shaft yet deeper in the mountain, and poured in even more tons of concrete. It also added security fences on the surface.

At last, however, after fifty years, when few were alive who had ever seen the great Buffalo Bill in person, tempers cooled. No one in Nebraska cared much anymore where the old hero was buried. In Wyoming, the folks of Cody finally gave up and formally ended their quarrel with Denver. In 1968, an end of hostilities was marked by an exchange of smoke signals between Lookout Mountain and Cedar Mountain, while the spirit of Buffalo Bill was transported symbolically from one mountain to the other on a riderless white horse. The people of Cody relinquished their claim to their founder's corpse, while the urbanized denizens of Denver tacitly admitted what everyone had known all along, that Cody's spirit belonged in Cedar Mountain.

Here Lies Sitting Bull

Tatanka Yotanka, or Sitting Bull in English, the staunch leader of the Hunkpapa Teton Sioux, won lasting fame as one of the chiefs who defeated General George Armstrong Custer at the Battle of the Little Big Horn in 1876. Although he won the battle, he lost the war, and finally agreed to take his people to a reservation.

Buffalo Bill Cody was Sitting Bull's enemy during the war. When hostilities broke out, Cody had forsaken his fledgling career in show business to rejoin the U.S. Cavalry as a scout, and in a minor skirmish soon after Little Big Horn, he killed a Sioux warrior. This incident was magnified out of all proportion. Bill bragged that he had "taken the first scalp for Custer," and thereafter his theatrical fortunes improved immensely. By 1883, he had started his fabulous Wild West show. Looking for crowd-pleasing attractions, he invited his old antagonist Sitting Bull to join the troupe. The chief went on the road for a year, in 1885; he proved quite a draw, but he disliked the boos and catcalls that sometimes greeted him when he was introduced as the man who defeated General Custer. When Sitting Bull left the show to return to the reservation, Buffalo Bill, who was one of the few whites Sitting Bull ever liked or trusted, gave him a handsome gray trick horse as a token of their friendship.

Sitting Bull always knew he could not be happy on the reservation, cut off from the roving life he was born to, surrounded by troops, and dependent on the white man's subsidies. When messianic Ghost Dance mysticism swept the reservations in 1890, promising the demoralized Native Americans a magical return to the good old days, with plenty of buffalo and no white men, Sitting Bull listened, as did many of his band. He allowed his people to participate in the magical dance, which they hoped would restore unity, prosperity, and power to the Indians. The government authorities feared that the still potent chief was planning to lead his discontented faction off the reservation, with great potential for a war breaking out. The army asked Buffalo Bill to speak with Sitting Bull to defuse the situation, and he agreed, but the civilian Indian agent refused to allow him on the reservation. On the morning of December 15, 1890, James McLaughlin, the agent, sent forty-three of his Indian police to arrest Sitting Bull and bring him in for questioning.

As the Indian officers tried to take Sitting Bull into custody at his cabin on Grand River, in South Dakota, he stalled while 150 of his supporters gathered in the gloomy dawn outside. Playing for time, he insisted on riding the gray horse Buffalo Bill had given him. Just as it was brought, a shot rang out, a melee erupted, and in a moment Sitting Bull and twelve others lay dead. The outnumbered and surrounded police might have been exterminated if the gray horse had not suddenly reared up on its hind legs, as it had been trained to do at the sound of blanks in the Wild West show, and pawed the air in its best theatrical pirouettes. This unnerved the attackers, who fell back in confusion, giving the surviving police officers time to take shelter in the cabin. One of them jumped on the gray horse and raced for help.

The cavalry soon arrived to restore order. They brought an ambulance and a wagon, in which they placed the dead and wounded policemen. Despite the tears, pleas, and threats of Sitting Bull's wives and friends, they piled his body on the wagon too, and carried it forty-five miles north to Fort Yates, North Dakota. There Tatanka Yotanka was buried without ceremony in the post cemetery, sewed up in canvas inside a simple wooden coffin.

Clarence Grey Eagle, a Sioux youth of sixteen summers, had witnessed the death of his uncle, Sitting Bull. When he heard how the chief's corpse had been dumped in a hole like a dog's carcass, he knew it was not right. Three years later, he accompanied a delegation of Sitting Bull's relatives to Fort Yates to ask for the body, so they could rebury it among their own people with proper rites and ceremonies. Their request was refused. In 1908, when Fort Yates was decommissioned and the bodies of deceased soldiers and the slain Indian policemen were moved to another army post, the corpse of Sitting Bull was left behind in the deserted prairie cemetery, marked only by a concrete slab. Grey Eagle visited the forlorn spot every year, and kept trying to get permission to reclaim his uncle's remains from neglect and oblivion. Meanwhile, some of the white people in Bismark, North Dakota, concocted a plan to move the famous chief's body to their city, in hopes of attracting tourists. With the help of the South Dakota Historical Society, Grey Eagle was able to thwart this plan, but a bill introduced in Congress to provide for moving Sitting Bull back to his Grand River home in South Dakota and build him a monument was never enacted.

At various times Grey Eagle appealed to the superintendent of the reservation, to the government of North Dakota, to anyone he thought could help him move Sitting Bull's bones or at least take proper care of their lonely burial place. There was much talk, but never any action. At last, in early 1953, the Army Corps of Engineers announced that flood waters from its new Missouri River dam at Pierre would inundate Sitting Bull's grave. Grey Eagle knew that something had to be done, now or never. He enlisted the aid of his friend Walter Tuntland, president of the chamber of commerce in Mobridge, South Dakota, the town nearest to where Sitting Bull had lived. While Grey Eagle was busy tracking down all of Sitting Bull's descendants to get the legally required signatures authorizing exhumation, Tuntland persuaded the citizens of Mobridge that burying Sitting Bull in their neighborhood would be the right thing to do, and would be good for tourism, too. They formed an association that pledged $15,000 for a portrait monument by the sculptor Korczak Ziolkowski to be erected over Sitting Bull's new grave. There was one small problem, however. The state of North Dakota, having shown nothing but loathing for Sitting Bull while he was alive and nothing but neglect and contempt for his grave after his death, absolutely refused to permit the chief's relatives to transfer his body to another state, even though Grey Eagle had obtained all the necessary signatures and powers of attorney required by law and had gone through all the proper legal and bureaucratic procedures. The state health officer, Dr. Saxvik, refused to sign the disinterment permit and passed the buck to the state historian, who not only denied the request but immediately alerted the governor and other state officials, and telegraphed the state's congressmen that South Dakota was trying to steal one of North Dakota's valuable historical assets. Representative Otto Krueger protested to the Department of the Interior, which has jurisdiction over Indian reservations. Meanwhile Grey Eagle, Tuntland, and their allies rushed to secure the support of the government of South Dakota. The governors of North and South Dakota exchanged angry letters, and the war between the states began.

Newspapers on both sides of the state line began to cover the controversy, which was picked up and reported by journalists throughout the country and even overseas. As tempers rose, so did the water backing up behind the dam. It was clear that Sitting Bull's grave must be moved somewhere, or it would be at the bottom of a lake. Towns on both sides of the border, sensing from the publicity that there was potential money and fame to be had in claiming the Sioux leader's remains, began to compete for the honor of reburying them. Rapid City, McLaughlin, Bullhead, and Sturgis, South Dakota, all disputed the claim of their sister city Mobridge to be the final resting place of Sitting Bull. Bismark and Fort Yates in North Dakota also put in their bids. Some of Sitting Bull's relatives were enticed with promises of money to recant their signatures on the original reburial petition, in favor of other importunate towns. However, Grey Eagle convinced the waverers that nothing at all could be accomplished unless the heirs remained unanimous. The sculptor Ziolkowski, observing the unseemly squabbles among the states and municipalities, was reminded of the numerous Greek towns that pretended to be the burial place of the poet Homer, and aptly quoted the line from Thomas Seward's *On Homer*: "Seven cities claim the mighty Homer dead, where once poor Homer begged for bread." Meanwhile, rumors proliferated, adding an even more farcical atmosphere to the dispute. It was whispered that Sitting Bull's body was not in its grave, having been dissected long ago for medical research; that a Bismark entrepreneur had tried to obtain the chief's skin, to cut it up and sell the pieces as souvenirs; and that the head had been sent to the Smithsonian Institution in Washington, D.C., years before. When contacted by a lawyer for the Sioux and asked to return the skull, the Smithsonian trustees truthfully denied that they ever had it.

Both states appealed to various departments of the federal government. Since the original burial site was on a military compound within an Indian reservation in North Dakota, and since the proposed burial site was on the same reservation, but within the territorial boundaries of South Dakota, the question of jurisdiction was quite complex. North Dakota, to avoid the appearance of violating its own laws, while still refusing permission to remove the body, claimed that the decision rested with the federal authorities. Yet, when presented with letters from the military, from the Bureau of Indian Affairs, and the Department of the Interior, none of which objected to the heirs' wish to move the body, the governor of North Dakota still refused to act.

Grey Eagle was disgusted, though not surprised, at all the uproar. It was what he always experienced with white people whenever he raised the subject of his uncle's bones: lots of talk, plenty of tricks, but no action. Time was running out. Grey Eagle was an old man now, and the water in the new lake was rising fast. Although Grey Eagle did not understand the white man's language or his law, he knew what was right, and he knew that possession is nine-tenths of the law in anybody's legal code. Figuring that it would be easier to get forgiveness than permission, he decided to take matters into his own hands. On March 31, 1953, he and some friends went to Sitting Bull's grave to get the body themselves. On this first foray they were prevented by the superintendent of the reservation, but they regrouped, got better organized, and tried again in a snowstorm at dawn on April 8.

While one team dug a new grave at the proposed site five miles east of Mobridge, Grey Eagle with the town mortician, his hearse, and a truck full of diggers raided the abandoned cemetery where Sitting Bull reposed. The truck dragged the concrete slab away, and in a half-hour of furious digging the Indians uncovered all that was left of their ancestor's body, about one-third of his skeleton (including the skull). This time the superintendent interposed no objection, having received a telegram from the Department of the Interior telling him not to oppose the wishes of Sitting Bull's descendants.

In another hour the raiding party was at the newly prepared grave, where their compatriots waited with a cement truck and a pile of steel rails. Like his friend Buffalo Bill, Sitting Bull was now buried deeply under tons of concrete and steel, to settle the question of burial location once and for all. To make sure, the Indians stood guard for as long as it took for the concrete to solidify. The North Dakotans were outraged to find that Sitting Bull had been snatched away by a posse of South Dakota vigilantes. The newspapers called them ghouls, vandals, thieves, and scalawags. The governor fumed: "It was an underhand deal all the way through." He ordered an investigation by the state attorney general, who fulminated in vain that the South Dakotans had violated public health laws and other statutes. The South Dakotans replied by mockingly inviting the governor and attorney general to attend the grand dedication ceremonies for Sitting Bull's grave and monument on September 2. They both declined.

The state of South Dakota and the town of Mobridge went all out to make the dedication a memorable event. There were bands, parades, speeches, and five thousand native Americans from six states performing tribal games, dances, and rituals. Everyone was happy, everyone was satisfied, everyone felt triumphant. Tatanka Yotanka was home at last, honored as he deserved, thanks to the sixty-three-year persistence of his loyal nephew, Clarence Grey Eagle.

Ironically, the Army Corps of Engineers had miscalculated. The water behind its new dam never quite rose high enough to cover Sitting Bull's original grave.

Abraham Lincoln:
Ill at Ease in Illinois

One of the weirdest things that ever happened to a deceased American president was the attempt by some desperate counterfeiters to steal Abraham Lincoln's body and hold it for ransom. The price that the counterfeiters planned to ask for restoring the corpse was the release of one of their gang members from prison. Luckily, the Secret Service foiled the plot at the last second, so history never got the chance to record whether the authorities would have traded a live convict for a dead president.

This crazy scheme to steal Lincoln's corpse was only the most dramatic of many interesting experiences that befell the remains of the sixteenth president of the United States *after the funeral.* Indeed, Lincoln's body had a more exciting posthumous history than that of any other president: it was taken by special train on a twelve-day, 1,700-mile funeral journey, during which it was viewed by more than a million people; it popularized the new process of intravenous embalming; parts of it are preserved in a museum in Washington, D.C.; and the rest of it was moved or reburied seventeen times in thirty-six years before coming to rest for good (let's hope) beneath two tons of steel and concrete in 1901. Before narrating all these novel episodes in the career of Lincoln's corpse, however, we should refresh our memory about how Lincoln came to be a corpse.

As everybody knows, Abraham Lincoln was born in a log cabin in Kentucky. He grew up on a farm in Illinois, taught himself law while splitting fence rails and clerking at a country store, and was so upright and trustworthy that he was still known as Honest Abe even after he became a lawyer and a politician. He was a gifted orator, a natural storyteller, and an antislavery crusader. He was elected president in 1860, but he was so unpopular in the southern slave states that for his inauguration in 1861, he had to sneak through Baltimore and into Washington, D.C., to avoid being assassinated. His election helped precipitate the secession of half the states, followed by the bloody Civil War. As Union armies lost battle after battle in the early years of the conflict, Lincoln became almost as unpopular in the North as he was in the Confederacy; but his perseverance and fortitude finally resulted in solid victories for the federal forces, emancipation of the slaves, and the start of a process of national reunification and reconciliation. He was now hailed as the savior of the nation, and was easily elected to a new term of office.

On April 14, 1865, shortly after his second inauguration, Lincoln was assassinated by John Wilkes Booth, an actor who crept into his box at Ford's Theater in Washington and shot him in the back of the head. The unconscious victim was carried across the street to a private house, where doctors tried in vain to save his life. He died early the next morning.

Dr. Edward Curtis and others performed an autopsy on Lincoln at noon, April 15, about four hours after he died, but since the cause of death was obvious, this was mostly a formality. About one gram of skull fragments, seven tiny pieces in all, and some bloody bandages, plus Dr. Curtis's bloody cuff, were saved after the procedure and eventually found their way into the National Museum of Health and Medicine (originally the Army Medical Museum) in Washington, D.C., where they are still on display. More about them later.

The nation was stunned by the assassination of the president. Outpourings of grief and mourning took place everywhere north of the Mason-Dixon line, and most white southerners were aghast at the deed, while blacks everywhere treated the loss of their great benefactor as a personal tragedy. Public tributes and memorials were held in towns and villages across the land. Congressional leaders wanted to bury the martyred leader in the ceremonial vault under the rotunda of the Capitol. This crypt had been built for George Washington, who was practically a saint in the eyes of all Americans, but George's descendants would not violate their famous ancestor's dying wish that he be buried at home. Thus, his place in the Capitol had been embarrassingly vacant for decades. Now Congress was spurned again. Its generous offer to inter Abraham Lincoln in the hallowed sepulcher was politely declined by his wife and son, who, at the advice of friends, decided to take Abe back to Springfield, Illinois, the scene of his early life and triumphs.

Congress and the leaders of the executive branch then agreed that the next best thing to burial in the Capitol—better, in fact—would be a solemn funeral train to carry the body back to Illinois, with public obsequies and ceremonies at major cities all along the way.

Fortunately, at the time of Lincoln's murder, Washington, D.C., was the center of the nascent embalming movement. The modern technique of embalming by injecting chemicals into the circulatory system was first patented in the United States in 1856 by a resident of Washington, D.C., J. Anthony Gaussardia. Soon, many other patents for rival forms of fluid embalming were issued, most of them to Washington inventors. Although the army had adopted this idea as the only practical way of preserving thousands of dead soldiers for shipment to their faraway homes, the newfangled method of embalming had not yet become popular with the public at large. Suddenly, however, the preservation of Honest Abe gave it a million dollars worth of free publicity. Lincoln's corpse was subjected to the new treatment, which enabled it to withstand the two-week trip back to Illinois in great shape, with only minor deterioration. Along the way, over a million people approvingly viewed the lifelike corpse; and the new method of embalming, as well as the practice of keeping embalmed bodies on display for days, came into a vogue that has not yet run its course.

There was some delay in getting ready for Lincoln's big send-off, because the preparations were extensive and the government was in disarray after the assassination. Finally, on April 18, five days after the murder, the body of Abraham Lincoln lay in state in an open mahogany casket at the White House, where twenty-five thousand people viewed it. The following day, after a restricted-access funeral service, a military parade two hours long escorted the bier through throngs of spectators from the presidential mansion to the Capitol. Here the body was placed in the rotunda for public viewing all the next day, with suitable speeches and other observances.

A day later, April 21, a cortege conveyed the coffin to the Baltimore and Ohio Railroad depot, where a doleful special train of eight coaches had been assembled. The heavy sixteen-wheeled presidential car, in which the commander in chief used to visit his armies on the front lines, now bore the deceased leader one last time. It also carried the small body of Lincoln's son Willie, whose death three years earlier almost caused Lincoln to sink into despair. Now the boy was lovingly exhumed to accompany his deceased father and grieving mother home to Illinois, tracing in reverse the journey Lincoln had made to Washington four fateful years before.

The first stop was Baltimore, where his reception was far more cordial than it had been in 1861, when he was widely detested and in danger of his life. Now he was eulogized, praised, and lamented. Thousands came to see his corpse and mourn their loss. The same scene was repeated in Harrisburg, Philadelphia, and New York, where he was honored by the largest parade in the city's history. Then it was on to Albany, Buffalo, Cleveland, Columbus, Cincinnati, Indianapolis, and Chicago.

Everywhere the crowds came to pay their respects and listen to fulsome oratory. In villages where the train did not stop, the populace gathered from miles around, hats off and heads bowed, just to watch it pass. Farmers paused in their fields as the funeral train rolled by. At last, on May 3, it pulled into Springfield, the end of its twelve-day odyssey. Lincoln was back home.

Mary Lincoln had declined the offer to bury her husband in Chicago, as well as the plans of the governor of Illinois and the city fathers of Springfield to inter their favorite son in the center of the capital city. She insisted on putting her dear Abraham to eternal rest in the new garden cemetery, Oak Ridge, on the edge of town. His body was escorted by solemn parade to the cemetery on May 5 and, after a final ceremony, was deposited under armed guard in a temporary vault, until a more fitting mausoleum could be built.

They just couldn't let Honest Abe rest after that. The body was moved at least seventeen times and the coffin opened for identification five times during the next thirty-six years, although it never left the precincts of the cemetery. The first move, and the first identification, occurred in December 1865, when Lincoln was transferred to another specially constructed vault pending the erection of the National Lincoln Monument. This monument took nine years to complete, and was jinxed from the start. Lincoln was reidentified and reboxed in an iron coffin in 1871 and moved to yet another temporary site, within the partially erected monument, while the rest of the edifice, including a stone sarcophagus intended as his final resting place, was slowly finished. Only when flummoxed officials finally tried to insert the iron coffin into the sarcophagus three years later did they realize there had been a wee mistake in calculation: the sarcophagus was too small to hold the coffin! The only solution was to transfer the body again to a new, smaller coffin, this one of red cedar lined with lead, thus necessitating yet another formal identification. Finally, on October 15, 1874, President Grant formally dedicated the National Lincoln Monument, whose construction had lasted twice as long as the Civil War. Lincoln was laid to rest, supposedly forever, in the sarcophagus in the center of the Catacomb, the crypt of the Lincoln family.

Lincoln didn't rest very long. The next threat to his tranquility originated a year later, when master engraver Benjamin Boyd entered the state prison at Joliet, Illinois, for a ten-year enforced vacation from his job making counterfeit plates of U.S. currency. His boss, "Big Jim" Kinelly (or Kinealy) of St. Louis, was distraught, for although he could get all the printers and pushers he needed for his funny money, the Secret Service had confiscated all of Boyd's engraved plates, which Kinelly couldn't replace. He had to spring Boyd from prison before his inventory of phony greenbacks was depleted, for only Boyd could engrave plates to Kinelly's exacting standards: i.e., indistinguishable from the real thing. Counterfeit notes printed from Boyd's plates were so good, in fact, that the Treasury once had been forced to withdraw all five-dollar bills from circulation and change their design to protect the public from Boyd's prolific fakes.

Kinelly remembered hearing about a foiled conspiracy in 1865 to steal Lincoln's body for political purposes. This gave him the far-fetched idea to steal it himself and hold it for ransom. The ransom would be the release of Ben Boyd from the penitentiary, plus $200,000 in cash for good measure. The audacity of the scheme was amazing, but its execution was less than brilliant.

In his first attempt, he trusted the wrong men. Ben Sheridan, Kinelly's agent in Lincoln, Illinois, was sent with four accomplices to open a saloon in Springfield as a front. The plan was to snatch the body on July 3, 1876, and hide it in a beer cellar. But Sheridan bragged about the escapade to a local madam, who warned the sheriff, and the hapless gang fled town to avoid arrest.

Kinelly washed his hands of these bunglers and cast around for more reliable help. Traveling to Chicago, he enlisted two more of his operatives, Terrence Mullen and Jack Hughes. This time they planned to make the heist during the night of November 7, 1876, which was election day. The good citizens of Springfield would be too intent on who was stealing the presidential election to worry about who might be stealing the president. Also, the many farm wagons returning home from the polls late at night would allow them ample cover for the wagon in which they intended to smuggle the stolen corpse out of town. They would haul it two hundred miles and bury it beneath the dunes of northern Indiana, where shifting sands would obliterate all traces of their excavations.

This daring plan leaked out, not once but twice. Someone warned Robert Todd Lincoln, the president's son, who had a law office in Chicago. Another tip was given to the Secret Service office in Indianapolis, which passed the information on to Patrick Tyrell, the agent in charge of the Chicago branch. He ordered Lewis C. Swegles, a petty crook and paid informer for the Secret Service, to infiltrate the gang and find out more about their nefarious plot. Swegles had been spying on Mullen and Hughes, on suspicion of counterfeiting, ever since their associate Boyd had been arrested. Now, at Mullen's Chicago saloon, Swegles had no trouble ingratiating himself into their confidence, since he had experience in grave robbing, a practice that was rife in that era to supply medical schools with cadavers.

Swegles kept Agent Tyrell fully informed about the progress of the conspirators. Since it would be hard to convict them on the basis of merely planning a crime, and since body snatching wasn't even a crime in Illinois, Tyrell wanted to catch the miscreants in the act and charge them with burglary. When the night train for Springfield rolled out of Chicago on November 6, Mullen and Hughes, in the front car with Swegles, were unaware that the last car was full of Secret Service men and detectives hired from Pinkerton's and other detective agencies.

The following night, the sleuths hid inside the labyrinthine halls of the National Lincoln Monument in total silence and stygian blackness. The criminals arrived three hours later. Swegles was supposed to signal the lawmen as soon as his companions had broken into the Catacomb, but he had trouble sneaking away from them to do it. Not until the ghouls had entered the crypt, pried open the sarcophagus, and begun to drag the coffin out was Swegles able to slip outside on the pretext of fetching the wagon. He made a detour in the dark, ran to the other side of the monument where Tyrell was concealed with his men, and gave the word. The detectives charged outside and dashed to the door of the crypt, firing one of their pistols by accident. They were shocked to find no burglars inside, and began running around in the night shooting at each other in confusion; only by luck were none of them killed. The bandits, who had warily decided to wait outside in the shadows for Swegles to return with the wagon, scampered away as soon as the shooting started.

The Secret Service and the vaunted Pinkertons were temporarily covered with ignominy for this election-night fiasco, and Chicago newspapers even alleged that the entire episode was faked for some sinister political reasons. However, when Mullen and Hughes were arrested in Chicago, charged, and convicted on Swegles's testimony, all such imputations were forgotten. The pair was sentenced to a year in jail for burglary. Kinelly was not implicated, but was imprisoned four years later for counterfeiting.

The attempt to steal Lincoln's body led to the next of many more shifts, burials, and reburials of the Great Emancipator. To foil any other aspiring thieves, his friends secretly removed his coffin from the repaired crypt a few weeks later and hid it in various places inside the monument. For eleven years, visitors to the Catacomb paid their respects to an empty sarcophagus.

By 1887, the National Lincoln Monument was already in need of repair and remodeling. Since the president's body had to be moved for this anyway, and since rumors persisted that the corpse was not really Lincoln's, the coffin was opened for another verification. Despite repairs, however, by 1900 the monument was so dilapidated that it had to be almost completely reconstructed; Abe and his whole family were then displaced once more, to a temporary hole in the yard. There they stayed beneath nine tons of stone for almost a year before being moved back inside in April 1901. Lincoln was again moved from his sarcophagus in September of that year, reidentified for the fifth time, then replaced in the sarcophagus. His son, however, who still worried that someone might try to steal his father's remains, arranged for the coffin to be encased in steel bars, sunk ten feet beneath the floor, and covered with tons of cement. So far, this has been Abe's permanent grave, and they didn't even need to move him out of the way for yet more structural renovations of his monument in 1931.

The only loose ends left of Lincoln now are the skull fragments and blood samples displayed in the medical museum in Washington. Even these haven't been left in peace. In 1989, a college professor proposed to extract genetic material from the fragments to test the hypothesis of medical mystery enthusiasts that Lincoln suffered from a congenital connective-tissue disease called Marfan's Syndrome. This suggestion caused considerable controversy over the rationale, the practicality, and even the ethics of such a step. An ethics panel at the National Museum of Health and Medicine approved the study in 1991, but the following year another committee recommended a lengthy delay until more could be learned about which gene causes Marfan's Syndrome. However, the media attention given to the proposal resulted in hundreds of calls and letters to the museum from people who mysteriously claimed to own other pieces of Lincoln's anatomy! All these claims, of course, can't be true, but the museum hopes to investigate a few of the more plausible ones. Maybe after that, Abraham Lincoln will finally be allowed to rest in peace.

Livingstone's Last Journey

D r. David Livingstone, the famous African explorer, missionary, and antislavery crusader, was buried with all the pomp and honor that imperial Britain could bestow. His body was laid to rest in Westminster Abbey on April 18, 1874, a national day of mourning. Yes, the long-lost Dr. Livingstone was home at last in England, but his heart remained in Africa—literally; also his entrails and other internal organs. They were buried in a tin box under a large mulva tree in the village of Chief Chitambo, where he had died almost a year before.

How the rest of Livingstone's corpse made it from central Africa, a thousand miles inland from Zanzibar, to Westminster Abbey is a truly remarkable story of courage, loyalty, and determination on the part of his devoted African followers. It would have been far easier for them to have buried his body, entire, in Chitambo's village, and no one would have questioned such an obvious decision; indeed, it was the only logical, the only practical choice. But to Susi, Chuma, and their associates it was a matter of principle, not convenience; so even though no one made them do it, and no one could rightly have expected it, they carried their departed leader's embalmed cadaver on their shoulders for almost nine months, to deliver it safely into the hands of British authorities.

Dr. David Livingstone's loyal followers carried his mummified corpse over a thousand miles to the African coast so it could be shipped back to England. (*Courtesy of Library of Congress*)

David Livingstone, born in Scotland in 1813, rose from factory worker to worldwide fame. By hard work he gained a university education open to few of his class. He joined the London Missionary Society in 1838, and upon completion of medical school in 1840 was greatly disappointed in his wish to join the China missions. Instead, the society sent him off to southern Africa where, after some early misadventures, including nearly being eaten by a lion, he finally started to get the hang of things.

Livingstone was not the average missionary. Although he considered missionary work his primary occupation, he soon decided that the primary occupation of a missionary was to be an explorer. The number of people he personally converted was small, but this did not discourage him. He saw his true role as that of a trailblazer, doing the strenuous work of opening the way for others; therefore, the real measure of his success was his incredible record of geographic, scientific, and hydrologic discoveries. He also became obsessed with suppressing the slave trade that was devastating the societies of Africa, and his fearless efforts hastened the demise of that odious blot on human history. Practically everywhere he went during thirty-three years of crisscrossing equatorial Africa, Livingstone was welcomed and respected by the local people.

On April 4, 1866, David Livingstone started out from the East African coast on his last trek into the interior, sponsored jointly by the Royal Geographical Society and the British government. One stated goal of the expedition, of which Livingstone was the only European, was to oppose the slave trade, but the second objective consumed most of the doctor's time and energy. This was his futile search for the nonexistent "fountains of the Nile" of which Herodotus had written two thousand years before.

The beginning of this ill-starred quest was inauspicious, and then things got worse. By January 1867, all but five of Livingstone's men had died, deserted, or been discharged. All the animals and medicine and most of the supplies were gone. Drought, famine, war, and slave traders posed constant obstacles. Livingstone himself was a victim of illness. For over three years the indomitable explorer traversed the African interior around Lake Tanganyika; his dispatches became ever less frequent and his whereabouts less certain, and rumors of his death started trickling back to the British on the coast. By October 1869, most people assumed that he must be dead.

They were wrong. That very month, the American newspaper tycoon James Gordon Bennet gambled that Livingstone was still alive and that Henry Morton Stanley, a reporter for the *New York Herald*, could find him. Against all odds, the gamble paid off. With no experience of Africa nor of exploration, but backed by Bennet's bottomless purse, Stanley not only found Livingstone (it took more than two years), but lavishly resupplied him with men, provisions, medicine, and money just when he had reached his last gasp.

Stanley was supposed to rescue Dr. Livingstone and bring him back alive. The fly in the ointment was that Livingstone refused to be rescued. He had invested six years of his life searching for the fountains of the Nile, and he was determined to keep on and find them, or die trying. He died trying on May 1, 1873, at Chitambo's village in what is now Zambia.

Livingstone's entourage, loyal to the last through countless hardships, had refused to desert him, even though they knew he was dying. Even when they had to carry him, even when he was so weak he could not move, they cooperated with his every wish. Now the inevitable had occurred. The great leader was dead, and his men were stranded a thousand miles from home in a trackless continent rife with war, disease, and rapacious slavers.

Now they held a meeting. Susi and Chuma, who had been with Livingstone the longest, since 1866, were elected leaders. Jacob Wainwright, a missionary-educated African who had been one of the recruits sent back by the departing Stanley in 1872, was deputed to make a careful inventory of all the doctor's instruments, papers, and other possessions. Chitambo, the village chief, upon learning of his famous guest's demise, graciously waived the customary death fine usually imposed on transients for the bad luck that a death in their company could bring on his village. He also led his people in mourning the deceased, and helped the sixty expedition members prepare for their journey.

They needed all the help they could get. Well provided with trade goods, but short of food and medicine, it would be hard enough for them to reach Zanzibar without the impediment of taking Livingstone's body with them. Chitambo tried to talk some sense into them, and advised them to bury their lamented leader where he lay. He no doubt reminded them that given the morbid dread of corpses prevailing in that part of Africa, few villages would welcome a caravan bearing a cadaver, whose ghost would naturally be tagging along.

But Livingstone's men were adamant. Whatever the obstacles, whatever the risks, their duty was to see their leader returned for burial among his own people. They may incidentally have hoped for some reward, and they may also have been concerned to prove that they had not deserted Livingstone on a mere pretense of his death, as others had done before. But their primary motives were undoubtedly a mixture of love, respect, and devotion to duty.

Fortunately, one of the men had been a physician's attendant in Zanzibar and had some experience of autopsies. He advised his associates on the best way of preserving the body. After the men eviscerated the corpse, the heart and other organs were buried with proper religious rites under the mulva tree, upon which Wainwright carved the doctor's name. After the burial, Wainwright read the burial service from Livingstone's prayer book; they fired guns in the air and then, as one of the men later recalled, "we sat down and cried."

With Chitambo's permission, the party constructed a stockade near the village and spent fourteen precious days preparing the body for transport. Inside a specially built hut, open to the sky but with strong walls to deter animals, they packed the inside of the body with salt, bathed the face with medicinal brandy, and exposed the corpse to the desiccating rays of the tropical sun. The body was turned a fraction each day for uniform drying. Then it was wrapped in calico, encased in a bark cylinder, and sewed up in sailcloth, the arms and legs having been folded to make the result look as much like a bundle of trade goods, and as little like a corpse, as possible. Susi tarred the finished product to make it waterproof; then they lashed it to a carrying pole and set out for the coast.

Susi was an able and resourceful leader, but not long after starting he and half the men were incapacitated with fever. Despite losing a month to recuperate, they made remarkable progress, reaching Unyanyembe in five months. Along the way, news of Livingstone's death preceded them, and many people came to pay their respects; but many villages, out of superstition, refused admittance to the bearers of a dead body. In one district they came into conflict with the chief and had to fight their way out. They had lost ten men to hardship, violence, and disease by the time they reached Unyanyembe.

Here they met a belated expedition on its way up from the coast to rescue Livingstone again. Lieutenant Verney Lovett Cameron, the leader of this inept attempt, peremptorily appropriated all of Livingstone's scientific instruments for his own use, but Chuma and Susi absolutely refused to surrender the body to the overbearing English officer for immediate burial. They were determined to deliver it to the coast. This took four more months of toil, but they finally struggled into Bagamoyo in February 1874. Chuma hurried across the strait to Zanzibar to notify the authorities of their arrival.

John Kirk, the acting British consul at Zanzibar, was on leave, and Captain W. F. Prideaux was temporarily in charge. He sent a warship to Bagamoyo to pick up Livingstone's corpse from these dedicated men who, he was aware, had performed an incredible feat in bringing it there. But, lacking instructions and money, he was at a loss what to do for them. So he paid them their wages out of his own pocket and discharged them. He didn't even offer to bring them over to Zanzibar.

Livingstone's body was sent by military transport to Aden, then by steamship to Southampton, arriving on April 15, 1874. The Missionary Society paid for Jacob Wainwright's passage, but Susi and Chuma were left behind. James Young, one of Livingstone's financial backers, generously brought them to England the following year, where they finally received the honor they deserved. They also helped the editors of Livingstone's journals and papers, which they had preserved from oblivion. The rest of their brave party had long ago dispersed, and few ever learned about, much less received, the commemorative medals struck for them by the Royal Geographical Society. They received nothing from the British government.

When the P. & O. Lines steamer *Malwa* carried Livingstone's flag-draped coffin on its deck into Southampton harbor on April 15, it was greeted with a funeral dirge and a twenty-one-gun salute. A special train whisked the body to London where, at the headquarters of the Royal Geographical Society, a final identification was made. The face was unrecognizable, but the wounds to the left arm which Livingstone had sustained three decades earlier, when the lion almost had him for lunch, were conclusive. After lying in state in the society's rooms for two days, David Livingstone, minus his heart and other organs, was buried with great honor in Westminster Abbey. Thanks to the incredible self-sacrifice of Susi, Chuma, Wainwright, and his other followers, his last journey was at an end.

Eva Perón's Body Politic

E va Perón, better known as Evita, was the charismatic wife of Argentine president Juan Perón. Her maiden name was Duarte. She emerged from a mediocre career as an actress to become the galvanizing spirit, if not actually the power, behind the throne of her autocratic husband from 1946, when he first became president, until her death in 1952. She was a mixture of heroine, saint, benefactress, and spokesperson for the millions of poor Argentines, the "shirtless ones," on whom Perón's power rested. They idolized her as she simultaneously flaunted her wealth in furs and jewels while castigating with vehemence the rich and aristocratic enemies of her husband's regime. The upper classes, on the other hand, as they saw their privileges relentlessly attacked by the Perónists, came to resent and detest her; but to her adoring partisans she could do no wrong. Then she died.

Her cancer had been hidden from the people as long as possible, but finally she became too ill for the deception to go on. She died on July 26, 1952, at age thirty-three. Juan Perón was devastated by her death: first, because he really loved her; second, because he had come to depend on her; and last but not least, because he sensed that the strength of his position depended far more on her mesmerizing ascendancy over the

Evita Perón's perfectly embalmed body became a pawn in Argentine politics. (*Courtesy of Library of Congress*)

"shirtless ones" than on anything he did or said. They respected him; they adored her.

These considerations led him to take a radical step. Even

before Evita succumbed to her disease, he held private consultations with Dr. Pedro Ara about the possibility of having her body preserved. Dr. Ara was a professor of anatomy, a pathologist, a cultural attaché at the Spanish embassy, and possibly the greatest embalmer of all time. He disliked being called simply an embalmer, considering himself a practitioner of the "art of death." With his painstaking approach to mortuary science, dead bodies took on the grace and beauty of fine funerary statues; and his modest claim that they would never decompose seems close to the truth. He frequently carried a perfectly preserved and lifelike head of an old peasant in his luggage, as a testimonial to his skill. Yet he was discreet about his very expensive services, available only to the rich. Perón engaged him to immortalize Evita's body at any price. He intended to use it as a focal point for continued support of his rule, making it an object of pilgrimage for the common people.

The Argentines have a fascination, almost a political fixation, with the dead bodies of their leaders. In Argentina, political corpses are political capital, and live politicians are expected to make the most of them. In the previous century, for instance, General Juan la Valle's body was exhumed by his friends, disarticulated, and carried across the Andes in saddlebags for safekeeping in Bolivia. And then there was Juan Manuel de Rosas, who united Argentina in 1835, died in exile, and was buried in Southampton, England; pro- and anti-Rosas factions quarreled for a century over whether to repatriate his body to Argentina, and it was finally brought home amid great excitement in 1989. So, in Argentina, it was quite natural that Perón would think of putting the corpse of his beloved wife on display.

Perón ushered Dr. Ara into Evita's sick room almost as soon as she was dead, for the success of Ara's technique depended on getting started immediately. The first task, a preliminary step only, was to get the corpse into a condition to withstand a few days of official lying in state. The next day, Evita was moved to the Ministry of Labor, where she had wielded the reins of power for six years, through the trade unions which supported her husband's government. Here the people were allowed inside to view the body.

Even the Perónists were amazed at the vast, unchecked outpouring of grief for the glamorous, idolized Evita. The public viewing, supposed to last only a few days, stretched to a fortnight as the crowds kept coming. Over two million people filed past the glass-topped coffin, often in tears, having waited up to fifteen hours in rain and cold for a glimpse of their departed patroness. This could have gone on indefinitely, for the sorrow of the people was genuine and personal; however, Dr. Ara warned that unless he quickly started on the main part of the embalming process, the body could undergo irreparable deterioration. Perón abruptly stopped the public display; after all, he intended to put his wife on permanent exhibition inside a colossal statue of her, larger than the Statue of Liberty, in a park in downtown Buenos Aires; then no one would be deprived of seeing her. But first, Ara had to make her body imperishable.

After a final day of lying in state, this time at the Congress building, Evita was taken to the headquarters of the Confederation of Labor, or CGT. Dr. Ara was to finish his work there. He protested that it was unsuitable; a hospital or medical laboratory would be more appropriate. Perón insisted on the fortresslike office building, however, so he could better protect the body against reactionaries who might try to steal or desecrate it. He instantly converted a suite of rooms into the most modern, climate-controlled mortuary laboratory that money could buy, for money was something Perón had plenty of. He needed plenty, for Dr. Ara's meticulous processes of injecting and reinjecting the body with chemicals, submerging it in endless baths of acetate and potassium nitrate, and coating its surface with transparent layers of plastic, cost Perón over $100,000 and consumed a whole year. When the work was done, it was perfect. Evita looked better than life, and was virtually imperishable. The commission planning the giant mausoleum statue was so impressed that it asked Dr. Ara if he would consider treating four more corpses, to be used as mummified pallbearers for Evita's coffin inside her crypt. He declined; one corpse a year was all he could handle.

The construction of the megamausoleum never got beyond the stage of a massive hole in the ground (later converted into a public swimming pool) before Juan Perón was overthrown by a military coup in 1955. Perón was lucky to escape alive, and in his hasty departure he left Evita behind, still temporarily stored in a chapel-like room of CGT headquarters.

The military government systematically set out to efface all traces of Perónism, including the cult of Evita. All images of her were banned, on threat of imprisonment. However, when the generals took over the Labor headquarters and found Evita's actual body, they were perplexed as to what to do about it. Despite Dr. Ara's avowals, they could not quite believe that the lifelike figure was not just a clever effigy. He produced X rays. The generals suspected they were fraudulent and ordered new X rays. They were still skeptical. Finally, they cut off part of a finger to prove the body was a fake, and were nonplussed to prove exactly the opposite. This gave the military junta a serious problem. They had hoped it was only a statue, to be unceremoniously smashed and obliterated. A real body was something else. They wanted to "take the body outside of politics," so that it would never become a rallying point for the defeated but still numerous Perónists; yet they were reluctant, for religious reasons, to destroy it outright. The exiled Duarte family declined to take it; the Catholic Church would not grant an exception to its rules against cremation; Dr. Ara informed them that even if totally neglected, the body would never decay naturally; the Navy refused to bury it on its remote island base; and the new president, General Pedro Eugenio Aramburu, overruled a plan to dump it in the ocean from an air force plane.

The junta, after many anxious meetings, decided to bury Evita in an unmarked grave in a Buenos Aires cemetery, and clandestinely (or so they thought) removed her from the little chapel in the Labor building. Day after day, a military truck carried her furtively from one part of town to another, but somehow the Perónists always knew where she was, for flowers and candles would appear overnight whenever the truck was parked. The idea of a secret burial was abandoned, it being impossible to keep the secret. A cat-and-mouse game continued for two years, the junta trying to hide the body and the Perónists trying to find it. For a while, Evita's corpse was hidden in the apartment of an army major, until he accidentally shot his wife to death one night, mistaking her for a Perónist burglar. Next, Evita was stored in a wooden crate in the attic of military intelligence headquarters. Finally, in 1957, an elaborate hoax was devised, which resulted in the corpse being secretly sent, via the Argentine embassy in Bonn, to a graveyard in Milan, Italy, and buried under the name of Maria Maggi de Magistris, an Italian emigrant who had died in Argentina. The ever scrupulous Aramburu even arranged for masses to be said for the dead woman. As a precaution, identical coffins filled with ballast were sent to other Argentine embassies to throw Perónist spies off the trail.

For fourteen years Evita's whereabouts were known to only a few members of the junta. By 1970 Juan Perón, after many travels, had finally settled in Madrid with his third wife, Isabel. In Argentina, things had been going from bad to worse, both economically and socially. The chaos was getting so bad that even the military began looking to Juan Perón as the only leader who might be able to restore Argentina to a state of calm

Juan Perón kept his wife's embalmed body in his dining room. Later, his own corpse was desecrated, and the thieves demanded a huge ransom for his severed hands. (*Courtesy of Library of Congress*)

and order. Secret overtures resulted in Perón demanding certain preconditions, one of which was the restoration of his wife's body. By now, none of those in power knew exactly where her cadaver had been hidden. Former president Aramburu had left certain clues in a sealed envelope with his lawyer, with instructions to deliver it to the current president in the event of his own death. This eventuality conveniently came to pass in May 1970, when Aramburu was kidnapped by guerrillas and executed for alleged crimes against the people. The killers refused to give up his body until "the day the remains of our dear comrade Evita are returned to the people."

At this point the lawyer solved everybody's problem by producing Aramburu's sealed letter. After following the leads it contained, the government at last discovered the location of Evita's grave. In September 1971, with elaborate precautions of secrecy and the cooperation of the Italian, French, and Spanish governments, the body was found in Milan, exhumed, and rushed overland by hearse to Madrid without any apparent customs formalities. There it was delivered to Perón's villa; and in Argentina, Aramburu's corpse was returned to his family.

Dr. Ara, also living in Madrid, was summoned at once to Perón's residence. When the coffin was opened, the consummate artist of death was gratified to find Evita's body as fresh and lifelike as it had been when he last saw it in 1955. Dirt, a flattened nose, and a few cracks in the plastic coating, all easily remedied, were the only signs of wear and tear. There was nothing he could do, however, about the missing finger, whose sacrifice in 1955 had saved the corpse from destruction as a mere statue.

Perón had Evita's corpse cleaned and refurbished and kept it around the house. There are even eyewitness accounts by dinner guests that Evita's body was kept in the dining room where Juan and Isabel ate their evening meals. Isabel accepted her predecessor with good grace, even enthusiasm. However, when Perón returned briefly to Buenos Aires in 1972, and then permanently to reassume power in 1973, he unexpectedly left Evita behind in Madrid. This was not what the people wanted. When Perón died suddenly on July 1, 1974, he still hadn't got around to satisfying their wishes. Hereupon the same guerrillas whose kidnapping of General Aramburu four years earlier had led to the discovery of Evita's burial place now rekidnapped

Aramburu's corpse from the family vault and, holding it for ransom, demanded the return of Evita to Argentina. The blackmail worked, and the government flew her body to Buenos Aires in a chartered plane on November 17, 1974.

Now the feeble government headed by Perón's inexperienced wife Isabel began planning another grandiose mausoleum for Evita, and for Juan. Both bodies were kept on display in the presidential palace in the meantime. However, a military coup toppled the regime, and the new military junta at last, on October 22, 1976, handed over Evita's still immaculate corpse to the Duarte family. At government expense, a bombproof and (hopefully) burglarproof steel vault was sunk twenty feet beneath the family mausoleum in the fashionable Recoleta cemetery, the resting place of Argentina's rich and famous. Here Evita rests incongruously among the aristocrats of society whom she had so tirelessly opposed while alive. Her body is finally "out of politics"—at least for now.

Juan Perón's body was not so lucky. Shipped to Moscow for embalming (against his wishes) in 1974, it was returned to Argentina and buried in his grandfather's tomb in the Chacarita cemetery in suburban Buenos Aires. Despite elaborate security precautions, including twelve combination locks, bulletproof glass, and round-the-clock guards, thieves broke into his coffin in 1987 and stole his ceremonial sword. They also severed and stole his hands, for which they enigmatically demanded an $8 million ransom "for services rendered in 1972." The government fulminated, the Perónists went on strike and held mass rallies, and everyone enjoyed a great game of guessing who filched the hands, but since no one offered to raise any money to save them, the kidnappers presumably carried out their threat to "pulverize" them.

What next, Argentina?

CHAPTER 32

Why Jim Thorpe Is Buried in Jim Thorpe

J im Thorpe, the famous American Indian superstar of football, baseball, and the Olympics, died in California in 1953, three years after an Associated Press sportswriters' poll voted him the greatest athlete of the twentieth century (he received three times as many votes as Babe Ruth, who placed second). He is buried in the town of Jim Thorpe, Pennsylvania. This is no coincidence. The place was named after him. Not that it was his hometown or had any special connection with him: He never lived there, never went there, probably never knew It existed. Nonetheless, the citizens of Mauch Chunk and East Mauch Chunk enthusiastically merged their rival townships and formally adopted Jim Thorpe as their official civic name in exchange for the privilege of burying the renowned athlete's body among them.

Jim had no part in this bargain. He was already dead. His Sac and Fox kinfolk were not party to the deal either. They intended to inter him at the family plot in Shawnee, Oklahoma, until his widow unexpectedly drove away with his corpse. It was she who made the agreement that ultimately put Jim Thorpe's name on the map, and his body in the

Jim Thorpe, the man. (*The Artist's Proof, Inc.*)

ground of Carbon County, Pennsylvania. Here's how it happened.

Thorpe got his start when his sensational skill propelled the Carlisle Indian School football team to national promi-

Jim Thorpe, the town. (*Courtesy Pocono Mountains Vacation Bureau*)

nence. From 1907 to 1912, his team routinely trounced Ivy League opponents like Harvard, Yale, and Army. Jim also excelled in track and field, and he astounded the world by winning both the decathlon and the pentathlon at the Stockholm Olympics in 1912.

Although his gold medals were unfairly rescinded on a technicality, this humiliation did not prevent him from enjoying a distinguished career in both professional football and professional baseball, extending well into the 1920s. His later years, however, were not so bright. Fame never deserted him, but fortune did. He went from job to job, moved frequently, suffered family and health problems, and had some embarrassing incidents of fighting and other misdemeanors. He never did manage to capitalize financially on his fame and popularity and was generous to a fault: When he died in 1953 he was nearly broke, as usual.

His wife, the former Patricia Askew, had been Jim's tireless proponent in his quest for financial security. She had master-minded and managed the speeches, appearances, endorsements, and publicity campaigns of his last three years. When he died, it became her goal that her husband would be remembered in a way befitting the greatest athlete of the twentieth century.

She had no trouble convincing Jim's birthplace of Shawnee, Oklahoma, to bring his body home from California, but talk about a $100,000 memorial, a Jim Thorpe Motel, and a Jim Thorpe Cancer and Heart Foundation didn't seem to be leading anywhere. When the rent became overdue on the crypt where Thorpe's corpse temporarily resided, Patricia removed it to Tulsa. If Shawnee wouldn't grant her husband's remains the distinction they deserved, there were plenty of other towns that would. Or so Patricia thought. She diligently sought for a place that would make a suitable offer. But negotiations and feelers with various municipalities in Oklahoma, Ohio, and Pennsylvania came to nothing. Patricia eventually transported Jim's body to Pittsburgh, then Philadelphia.

It was in Philadelphia in 1954 that Jim's widow fortuitously saw a television report about the attempt of Mauch Chunk and East Mauch Chunk to revive their declining economies. They had scraped together some funds to foster industrial development through a "Nickel-a-Week" contribution drive sponsored by a local newspaper. Patricia Thorpe immediately contacted the town fathers and made them an audacious proposal. She would let them bury her husband's body if they built him a fitting memorial and named their dual township after him! She convinced them that Jim Thorpe's grave would be just the thing to attract thousands of tourists. Tourist spending for food, motels, and souvenirs would pump cash into the local economy, while the prestige of Jim Thorpe's name would aid the campaign to attract commercial development, starting with a projected Jim Thorpe Museum, a Jim

Thorpe Children's Recreation Center, and a Jim Thorpe Memorial Heart and Cancer Institute, built with donations from around the country.

The citizens of the twin boroughs were convinced. They enthusiastically agreed to bury Jim's body, as well as their long-standing rivalry, merge into one town, adopt the name of Jim Thorpe, Pennsylvania, and spend a big chunk of the Mauch Chunk development fund to construct a grand marble tomb. There were unsubstantiated allegations later that Mrs. Thorpe herself received $25,000 from the fund, with insinuations that she had sold her husband's relics to the highest bidder. These allegations are probably untrue. In any case, the townspeople were convinced that they had found the way to economic salvation.

It was not to be. Despite talk of a sporting goods manufacturing plant, a Professional Football Hall of Fame, and other projects, nothing much happened. The tourists stayed away in droves, too. Years later, a borough councilman lamented: "All we saw were dollar signs, but all we got was a dead Indian."

The town also got into a long-running dispute with some of Jim Thorpe's relatives, who wanted his body buried in Shawnee, Oklahoma, according to Indian tradition. This, they say, is what Jim wanted. He used to say that he was five-eighths Indian, three-eighths Potawatomie on his mother's side and one-quarter Sac and Fox on his father's side. (He was better at football than at arithmetic.) He is officially listed as one-eighth Indian by the Sac and Fox tribe to which he belonged, and he definitely was raised as an Indian and recognized as such. Some of his children think he should be reburied with traditional Sac and Fox obsequies, which had not been completed when Patricia Thorpe took his body away to Tulsa. They repeatedly have requested Jim Thorpe, Pennsylvania, to return their father's remains to Oklahoma, although they are reluctant to resort to litigation over what they consider a private affair.

The citizens of Jim Thorpe, although they profited but

Jim Thorpe, the tomb. (*Courtesy Pocono Mountains Vacation Bureau*)

poorly from their bargain for Jim's body, have so far declined to surrender their patronymic guest. Although he couldn't attract fortune for them any more than he could for himself, he did lend them a modest measure of the great fame he always enjoyed. And although the town never built the marble tomb it promised, the imposing red granite monument in its place displays the sincere words uttered to Jim Thorpe by King Gustave of Sweden at the Stockholm Olympics of 1912:

SIR, YOU ARE THE GREATEST ATHLETE IN THE WORLD.

PART IV

MISCELLANEOUS

A Word About Body Parts and Ashes

Heads, hearts, and whole bodies are not the full story when we consider the untoward accidents that happen to celebrated corpses *after the funeral*. An entire book could be written about the things that are done to people's miscellaneous body parts and their cremated ashes. A few such episodes have been dealt with incidentally in prior narratives, such as Voltaire's heel, Richard the Lionhearted's guts, and Thomas Hardy's ashes. The circumstances involving these celebrity body parts are far from unique. For example, after the emperor Napoleon Bonaparte died in exile at St. Helena, an autopsy was performed by four doctors. Although witnesses were present to prevent any unauthorized souvenir hunting, one of the surgeons secreted two small rib fragments, one of which still exists; he also kept two pieces of the ulcerated and perforated intestine (the probable cause of death), which ended up in the museum of the Royal College of Surgeons in England. Napoleon requested that his hair be cut off after death and given as mementos to various people, and these locks are still valued by collectors of Napoleana today, changing hands for hundreds of thousands of dollars. Strangest of all, a New York urologist claims to own a pickled piece of Na-

Death mask of Napoleon Bonaparte. (*Courtesy of Library of Congress*)

poleon's purloined penis, for which he paid "a substantial sum" in about 1972. Evidence for the authenticity of this little bit of history is meager, and most historians think the doctor was duped.

Napoleon was lucky. As we saw above, Louis XIV's heart was eaten by an English scientist. Even worse, part of the English king John's body was allegedly chopped up and used for fish bait. At various times, human skin has been used as fine leather for book covers, such as that of the mistress of the French writer Eugène Sue and several guillotined French aristocrats during the Revolution. The blood of other aristocrats of that era was used as pigment in oil paintings, some of which still hang in European museums.

On a lighter note, we find body parts just as often revered as reviled. Without even considering the thousands of fingers, teeth, toes, arms, and legs of saints kept as relics in Europe and Asia, we may briefly cite as one example the hands of Louis Braille, inventor of the touch reading system for the blind, whose hands are honored in Coupvray, France (his body is in the Panthéon of Paris). And Artemisia of ancient Halicarnassus, wife of King Mausolus, for whom she built the original Mausoleum, was so lonely after her husband's death that she mixed some of his ashes in her wine and drank them.

From hundreds of stories such as these about the unlikely fate of famous body parts and human ashes, I have selected the following half-dozen that I hope will be of most interest to my readers.

CHAPTER 34

The Tree That Ate Roger Williams

I f you visit the exhibit rooms of the Rhode Island Historical Society in the capital city of Providence, you can see almost everything left on earth of Roger Williams, founder of the third largest state in southern New England. What's left looks remarkably like a cartoon "stick figure" of a human body, including a back, legs, feet, an arm, and a hand. That's because it actually is a stick of wood, the root of an apple tree; but it used to be flesh and bones. It used to be Roger Williams. The story of how Roger Williams was transmogrified into an apple tree root is a bizarre example of the unusual things that can happen to a mortal person *after the funeral.*

Roger Williams, born in England about 1604, emigrated to the Massachusetts Bay Colony in 1631 in search of religious freedom. It took less than six months for his insistence on complete religious freedom of conscience to get him into trouble with the theocracy of Boston. By 1636 he had been tried as a dissident, convicted, and banished, so he and a few followers trekked south and founded their own colony, Providence Plantation, later expanded and renamed Rhode Island.

The colony flourished as people flocked there from England and the other colonies seeking religious tolerance. This idea, radical at the time, later became one of the pillars of

American freedom, enshrined in the United States Constitution and emulated throughout most of the world.

Roger lived to a ripe old age, dying at Providence in March 1684. There being no public cemetery, he was buried on his farm, on a hill behind the house. The grateful citizens of the flourishing colony he had founded forty-eight years earlier gave him a big funeral with full military and civil honors.

His wife, Mary, was later interred next to him. After that, his body rested peacefully until 1739. In that year, a gravedigger preparing to bury one of his descendants dug too closely and accidentally broke open the upper end of Roger's coffin. A certain Captain Asa Packard, at that time a ten-year-old boy, later noted this fact in his reminiscences, for he was impressed by being able to peer into the coffin of the legendary patriarch of the colony.

Perhaps it was because of this accident that the apple tree root first came into the coffin, and into this story. For 121 more years, no one disturbed the grave of Roger Williams. Then, in 1860, the city of Providence decided it had been remiss in honoring the resting place of this great man, which was now, after urbanization and subdivision of the original farm, located somewhere in the back yard of the Sullivan Dorr house near the corner of Bowen and Pratt streets.

The precise location of the gravesite being uncertain, subterranean investigations were authorized to enable the monument builders to erect their proposed tribute, a marble column, on exactly the right spot. The results of these excavations were told by J. Dorman Steel in his 1868 book, *A Fourteen Weeks Course in Chemistry*, to illustrate the biological interdependence of the vegetable and animal kingdoms.

For the purpose of erecting a suitable monument in memory of Roger Williams, the founder of Rhode Island, his private burying ground was searched for the graves of himself and wife. It was found that everything had passed into oblivion. The shape of the coffins could be traced only by a black line of carbonaceous matter. The rusted hinges and nails, and a round-wooden knot alone remained in one grave; while a single lock of braided hair was found in the other.

Near the graves stood an apple-tree. This had sent down two main roots into the very presence of the coffined dead. The larger root, pushing its way to the precise spot occupied by the skull of Roger Williams, had made a turn as if passing around it, and followed the direction of the backbone to the hips. Here it divided into two branches, sending one along each leg to the heel, when both turned upward to the toes.

One of these roots formed a slight crook at the knee, which made the whole bear a striking resemblance to the human form. There were the graves, but their occupants had disappeared; the bones even had vanished. There stood the thief; the guilty apple-tree, caught in the very act of robbery. The spoliation was complete.

The organic matter, the flesh, the bones of Roger Williams, had passed into an apple tree. The elements had been absorbed by the roots, transmuted into woody fibre, which could now be burned as fuel, or carved into ornaments; had bloomed into fragrant blossoms, which had delighted the eye of passersby, and scattered the sweetest perfume of spring; more than that, had been converted into luscious fruit, which, from year to year, had been gathered and eaten.

When first discovered *in situ*, the root system that filled the clearly outlined coffin space of Roger Williams must have had a much more dramatic resemblance to a human body than the unimpressive sticks on display at the Rhode Island Historical Society today. The smaller branches, subroots, and cilia (root hairs) that once vividly mimicked the body's circulatory system were unavoidably damaged in excavation and have since disappeared entirely, leaving only part of the main root, vaguely resembling the lower skeleton from the spine to the feet, plus one forearm and hand. It takes some imagination to be convinced that these fragments look more manlike than any other old sticks found in the woods. But that's not the point. The important fact is that these bits of woody root are, in part, the very substance of Roger Williams.

Since nothing identifiable as Roger Williams remained except the apple tree root, the excavating committee of 1860 scooped up some of the loam that surrounded it and, on the reasonable assumption that some molecules of Roger must be in there somewhere, deposited it in the Stephen Randall tomb in the North Burial Ground. Later it was put in a metal container and moved to the main cemetery vault. In the late 1930s, this container was at last transferred to the base of the Roger Williams statue on Prospect Terrace.

Finding Einstein's Brain

"I want you to find Einstein's brain." These were the words Steven Levy, a reporter for the magazine *New Jersey Monthly*, heard with surprise from his editor one day in 1978. "Sure," he answered confidently, although he did not even know the famous brain that had formulated the theory of relativity was missing, nor how much trouble it would be to track it down. His editor, who had done a little preliminary research, quickly explained the situation to his puzzled reporter. It seems that when Albert Einstein died in 1955, his brain was removed for study and had never been heard of again. The editor wondered why, and told Levy to find out. It was an assignment reminiscent of publisher James Gordon Bennet's dispatching his reporter Henry Morton Stanley to find Dr. Livingstone, except that Stanley knew Livingstone was somewhere in Africa, whereas Levy didn't know where on earth Einstein's brain might be.

Steven Levy, after a long investigation, eventually traced the lost brain (or most of it) to a couple of bottles of formaldehyde in a cardboard box, stashed beneath a beer cooler on the floor of a doctor's office in Wichita, Kansas. The doctor was Thomas Harvey, who had been the staff pathologist at Princeton Hospital in Princeton, New Jersey, when the famous scientist died there on April 18, 1955. During a routine au-

Albert Einstein's brain was lost for over twenty years. (*Courtesy of Library of Congress*)

topsy, performed in the presence of Dr. Otto Nathan, Einstein's friend and literary executor, Dr. Harvey removed the brain and put it aside in a jar. The rest of the body was taken briefly to a local funeral parlor for a private ceremony. Later that day it was cremated in Trenton. Since Einstein had been emphatic that he did not want his body to become a focal point for hero worship or a "personality cult," Dr. Nathan, in accordance with Einstein's wishes and those of his family, scattered the ashes in a river, probably the Delaware.

But why wasn't Einstein's brain cremated along with the rest of his body? Whether Einstein wanted or agreed to have his brain abstracted from his body and preserved for scientific study is still a debated issue, as is the question of whether Otto Nathan knew that the brain had not been cremated with the rest of the corpse. Everyone connected with the incident and with the subsequent history of the brain has maintained an almost impenetrable wall of secrecy, ambiguity, and obscurity, punctuated occasionally by flashes of contradiction and confusion. Who, if anyone, authorized Dr. Harvey to keep the brain? If any research had been done on it, why hadn't the results been reported? Why did Harvey still have most of the brain in his private possession twenty-three years later? And where was the rest of it? Hard answers to many of these questions are still elusive. It seems clear that embarrassing facts were and are being concealed by all the knowledgeable participants, under the general excuse that Einstein's wish for privacy must be honored. One possible interpretation of events, as they have been ferreted out by Levy and other inquisitive journalists through the years, is as follows.

Despite later claims to the contrary, Albert Einstein did not will his brain to science. However, his friend Dr. Harry Zimmerman was quoted in 1994 as saying that Einstein asked him personally to examine his brain when he died. The family and Nathan were probably aware of this, but assumed that some simple examination would take place as part of the autopsy, such as dissecting, weighing, photographing, measuring, and taking small blood and tissue samples, after which the brain itself would be cremated with the rest of the body. Nonetheless, with or without Nathan's knowledge, Dr. Harvey simply put the brain aside after the autopsy and quickly notified Dr. Zimmerman, for whom he had formerly worked.

Some sources indicate that Princeton Hospital, or at least its administrator, had asked Harvey to save the brain, with hopes of participating in the projected studies of it. In any case, Zimmerman immediately talked to the press. On April 20, only two days after the autopsy, the members of Einstein's family were startled to learn from an article in the *New York Times* that Albert's brain was still in existence and about to be subjected to well-publicized scientific scrutiny. In fact, more details were promised by Zimmerman in a press conference scheduled for the following week.

The press conference never took place. The family was outraged that such a step was taken without its permission and that the culprits were so eager to publicize their plans for the brain when it was clear that Einstein had done everything possible to avoid attracting attention to his dead body. However, the family was in a quandary. It was implied by the press that Einstein had donated his brain to science and that great discoveries lay in store for the researchers who could unlock the secrets of the greatest brain of the last three hundred years. Trying to back out of this position now would not only be embarrassing but would generate even more unwanted publicity.

A compromise was worked out. With the acquiescence of the family, Dr. Nathan made a statement at the reading of Einstein's will nine days later: "Permission was granted for a study of the tissues of the brain, but with the express and emphatic proviso that this be done, if at all, in strictest privacy, without announcement or publication of any kind, except that the findings might be made available to the scientific world through medical journals, and not otherwise."

This interdict was amazingly effective. Nothing more was published or broadcast about the brain in newspapers, books, magazines, radio, or television. The trouble was that nothing was published in any medical or scientific journal either. This black hole of information about the brain was what prompted the editor of the *New Jersey Monthly* to send Steven Levy on his quest for the brain twenty-three years later. Had any research been done? Was the brain still available for research? Or had it been lost?

Apparently Thomas Harvey, the pathologist of Princeton Hospital, had been given the task of coordinating research on the brain. Although he had no specialized training in the physiology or chemistry of brains, he appears to have been chosen by the hospital, which had the brain in its possession and wanted some credit for any discoveries that might result. Harvey was expected to work closely with Dr. Zimmerman and others who had better credentials in brain studies; Harvey's role would be that of custodian of the "gross matter," as they called it, and general research coordinator.

Initial consideration of the brain as a whole showed, as expected, nothing extraordinary about its size, shape, weight, or other physical characteristics. Harvey then duly prepared the famous brain by having most of it sectioned into two hundred exactly mapped, photographed, and cataloged pieces, some of which were encased in celloidin, and then everything was preserved in formaldehyde. Tissue samples of the various parts were prepared on microscope slides. Harvey next doled out sections and slides to researchers around the country, including Zimmerman and others, eight in all. He usually delivered the samples in person, and sometimes, as in the case of Dr. Sidney Schulman of Chicago, who analyzed portions of the thalamus, went back to retrieve the samples after the work was done.

But why was nothing published? Not until thirty years later, in 1985, did one small report appear, in the journal *Experimental Neurology*, and that was by researchers who obtained tissue samples from Dr. Harvey only after learning his whereabouts from Steven Levy's article about his successful search for Einstein's brain. Nothing else was published because nothing significant was learned; the scientific techniques of the

1950s were not as sophisticated as those developed later, such as DNA matching. Meanwhile, the preservation of the brain tissue in formaldehyde for so long had made it unsuitable for any later testing. As far as science can determine, there is nothing unique about the structure or chemistry of Einstein's brain to serve as an indicator of genius. It is apparently just another brain.

When Dr. Harvey left Princeton Hospital in 1960, scientific interest in Einstein's brain was already ebbing. Harvey took the brain parts and all pertinent records with him when he moved, first to Freehold, Kansas, and then on to Wichita, where he worked at a biological testing laboratory. It was in Wichita that Steven Levy caught up with him in 1978. Later that year he moved to Westport and opened a medical practice; on retirement, he moved to Leavenworth, then Lawrence, both in Kansas. When interviewed by a reporter for the *Wall Street Journal* in 1994, at age eighty-one, he was worried about what would become of the brain, of which he still had about 75 percent, when he died. In that year, he gave a small slice of it to Professor Kenji Sugimoto of Kinki University in Japan, who has published two books about Einstein and plans to house the bit of his brain in a projected Einstein Museum.

However, other museums were reluctant to accept the rest of Einstein's brain, of which Harvey had then been the caretaker for almost four decades. According to newspaper reports, the Hebrew University in Jerusalem, the Einstein College of Medicine in New York, the Wistar Institute in Philadelphia, and the Smithsonian Institution in Washington, D.C., all expressed lack of interest. Thus, the brain that revolutionized science is of no further scientific interest, not even as a museum exhibit. Alive, the brain was of inestimable value to the human race; dead, dissected, and preserved, it is useless; all of which goes to prove that like everything else in the universe according to Einstein, value is relative.

CHAPTER 36

Joe Hill's Ashes: Addressee Unknown, Return to Sender

I n 1986, an employee at the United States Archives made an unexpected discovery while sorting some boring Post Office records from 1917. It was an ounce of cremated human ashes. They were found in a small, torn envelope with an identifying photo of the deceased printed on the front, one of his poems on the back, and a card bearing precise disposal instructions.

The ashes were those of Joe Hill, the legendary labor radical, poet, and songwriter executed for murder in 1915. How a wee pinch of Joe Hill became buried in the National Archives is a tale well illustrative of the strange, unforeseen events that may afflict a person's mortal remains *after the funeral.*

In Joe Hill's case, it was after *two* funerals: two big, emotional, and radical funerals, one in Utah and the other in Chicago. But first, let's review the circumstances that led up to the misfortune of Joe's needing a funeral in the first place.

Joe Hill was born in Gävle, Sweden, in 1879. His original name was Joel Emmanuel Hägglund. Working mostly as a seaman, he learned English pretty well before emigrating to the United States in 1902. Upon arrival, he changed his name to Joseph Hillstrom, which he later shortened to plain Joe Hill.

He spent the next ten years as an itinerant farm laborer, builder, dockworker, seaman, and miner. As he traveled around the country, he became increasingly radical in his views. In 1910, he joined the Industrial Workers of the World, or IWW, whose members were known derisively as the "Wobblies" to their enemies—and they had plenty of enemies.

The Wobblies, founded in 1905, were extreme labor radicals, militants expecting the violent overthrow of the capitalist system and doing their utmost to hasten the day. Their fiery rhetoric and confrontational tactics, which rejected negotiating and the political process in favor of "direct action" such as strikes, boycotts, and slowdowns, alienated them from the more moderate mainstream of the American labor movement. The respectable middle and wealthy classes, naturally, saw the Wobblies as anarchists, revolutionaries, and criminals. So did the government. Still, despite its many enemies, the IWW had a small but dedicated following among unskilled workers, migrants, and immigrants not well represented in the conventional labor movement. It was especially strong among the restless proletariat of the western mining and transport industries.

After Joe hooked up with the Wobblies, his natural talents as an agitator, poet, and songwriter quickly made him a minor celebrity in the movement. As he traversed the West, from the San Pedro docks to the copper mines of Utah, job hopping and getting involved in workers' struggles, he kept sending his lyrics to IWW headquarters in Chicago. Many of these bitter parodies of popular songs were published in the Wobblies' famous *Little Red Songbook* and are still sung on picket lines today, such as "The Preacher and the Slave," "Casey Jones," and "The Rebel Girl." Although an immigrant, Joe had a keen ear for the American idiom, along with a gift for satire.

He also had a gift for organizing. In 1913, he helped promote a successful strike against the United Construction Company in Bingham, Utah. This did not endear him to the ruling

classes in the state. By January 14, 1914, however, Joe was "between jobs," living precariously in Salt Lake City, and not
holding any official position in the IWW.

It was on this fateful day that two masked men murdered
John G. Morrison and his son Arling in their Salt Lake City
grocery store. Arling shot one of his assailants before being
killed. When Joe Hill called at a nearby doctor's office an hour
later, suffering from a gunshot wound, he became the prime
suspect in the double homicide.

Hill steadfastly but unconvincingly maintained that he
was wounded by a rival in a dispute over a girl, but he refused
to identify either person or name any corroborating witnesses.
On the other hand, the prosecutor's case against Joe was weak
and circumstantial. The only witness, Morrison's surviving son
Merlin, could not positively identify Joe Hill as the murderer.
Police never found the murder weapon or any other evidence
linking Hill to the crime or to the Morrisons, whose killings
had been actuated not by any labor dispute, but probably by
simple revenge. John Morrison, a former policeman, had
many enemies and had survived two previous assassination
attempts.

However, the state authorities and the conservative newspapers pressed for a conviction of their only suspect, whipping
up public excitement about the case based on Hill's reputation
as an IWW activist. Clearly, there were bigger issues at stake
than a dead grocer. Hill represented anarchy, revolution, law-
lessness, and radical challenge to established authority. He per-
sonified the threat to true American values of uncontrolled
immigration. He was not even a Mormon, which was a great
disadvantage in the state of Utah.

As publicity surrounding the case mounted, the IWW
came to Joe's defense. To them it was obvious that Hill was
being railroaded and framed for murder, although his only real
"crime" was that of being a Wobbly who had made himself
obnoxious to the capitalist interests of the state.

Hill might have been guilty of the murder, but there certainly was not enough evidence presented to convict him "beyond the shadow of a doubt," as the law demanded. Yet convicted he was, and sentenced to die. After the verdict, the Industrial Workers of the World organized a massive public relations campaign to save him from execution. Its efforts, plus the notoriety of the case, resulted in over seventy-five thousand letters to the governor pleading for retrial or clemency.

The cause célèbre grew to international proportions. The Department of State received so many protests from abroad that it appealed to the state of Utah for a reprieve. The government of Sweden, Joe's homeland, demanded further investigation, as did President Woodrow Wilson, who twice telegraphed Governor Spry on Joe's behalf. All was in vain. Joe was executed by firing squad on November 19, 1915.

Joe Hill now became a martyr for the labor movement. Both he and the IWW had foreseen this eventuality and prepared for it. "What do I expect to accomplish by my situation?" Joe said before he died. "Well, it won't do the IWW any harm, and it won't do the state of Utah any good." To Big Bill Haywood, the flamboyant Wobbly president, Joe wrote on his last night "Goodbye Bill: I die like a true rebel. Don't waste any time mourning—organize!" Joe refused to allow his body to be given to the University of Utah medical school for dissection, and asked Bill, "Could you arrange to have my body hauled across the state line . . . ? I don't want to be found dead in Utah."

This was fine with Bill Haywood and the organization, who had great plans for Joe's body. Too valuable to the movement to be simply buried, it would be put to good use to further the cause. It would be cremated to ashes, then parceled out to IWW locals throughout the world, to be scattered to the winds simultaneously on May Day, 1916. Joe may have given the IWW this idea himself. His last poem, written from prison and entitled "My Last Will," reads:

My will is easy to decide
For there is nothing to divide
My kin don't need to fuss and moan
"Moss does not cling to a rolling stone."
My body? —Oh!— If I could choose
I would to ashes it reduce
And let the merry breezes blow
My dust to where some flowers grow.
Perhaps some fading flower then
Would come to life and bloom again.
This is my Last and Final Will.
Good luck to all of you.

Joe Hill

Capitalizing on the international excitement surrounding Hill's execution, the Wobblies first staged two elaborate public funerals for the dead hero. The first was in Salt Lake City on November 21 and, to the dismay of the Utah authorities, it attracted thousands of workers. Speakers eulogized Joe Hill as a scapegoat murdered by the capitalists. After the service, a solemn procession, featuring six white-robed women pallbearers, bore the body to the train to be hauled out of the abominated state of Utah.

It was hauled forthwith to IWW headquarters in Chicago, where an even bigger funeral was adroitly orchestrated by Big Bill Haywood for Thanksgiving Day, November 25. Thousands viewed the body on the previous afternoon, and five thousand mourners jammed the West Side Auditorium for the service itself, while the streets were filled for blocks around, stopping all trams and other traffic.

The crowd, inside and out, rambunctiously sang Joe Hill's best labor songs and listened to hours of impassioned harangues castigating the assassins of the innocent man and preaching the rights of labor and social revolution. They could hardly get the coffin through the crowds and into the hearse. A gigantic funeral cortege followed the hearse to the train station for the short ride to Graceland Cemetery. There, Irish labor leader James Larkin and others gave more excited speeches, and the surging, singing crowd didn't completely disperse until late that frosty night.

Joe's body was cremated at Graceland the next day. As planned, his ashes were put to use in spreading the gospel of proletarian revolution. The IWW printed small coin envelopes with the hero's photo on the front and his "last will" on the back. At its 1916 convention, these were distributed to its locals from every state except, of course, Utah, where Joe "didn't want to be found dead." Packets also went to South America, Europe, Asia, South Africa, Australia, and New Zealand.

The plan was for each local to release Joe's ashes to the winds on May 1, 1916. This was presumably accomplished in most cases, although one stray package turned up three years later in a Toledo union hall and was not duly scattered until 1950. A few other envelopes were left over, and though most were confiscated and destroyed after a government raid on IWW headquarters on September 5, 1917, one of them apparently escaped and ended up in the National Archives in Washington, D.C.

Prompted by the Great War in Europe, Congress had passed the Espionage Act of 1915, which prohibited sending seditious material through the mail. On October 8, 1917, one month after the raid on Wobbly headquarters, postal machinery in Chicago accidentally mutilated one of the packets bearing Joe Hill's seditious remains, exposing the contents. The package was addressed to a mysterious Charles Gepford, never identified, of Chicago.

The suspicions of the postmaster being aroused, he confiscated the packet and forwarded it to the solicitor of the Post Office Department in Washington, D.C. The solicitor duly notified the Bureau of Investigation (forerunner of the FBI) of his catch and filed it away waiting for a response, which never came.

In 1944, Post Office files pertaining to the Espionage Act were routinely transferred to the National Archives, whose employees didn't get around to classifying them until 1986. That's when Joe's ashes were brought to light, after burial in the bowels of the bureaucracy for almost seventy years.

In the spring of 1986, the National Archives mentioned the ashes in a list of unusual items in its possession. The ashes might well have remained in a government file if someone at the Potomac Labor History Society had not noticed their mention two years later in the list of archival curiosities. This so-

The front of the envelope in which Joe Hill's ashes were discovered in the U.S. National Archives, and the reverse of the envelope. (*Courtesy of National Archives and Records Administration*)

> FELLOW WORKER:
>
> In compliance with the last will of Joe Hill, his bod
> was cremated at Graceland Cemetery, Chicago, Illino
> Nov 20th, 1915.
>
> It was his request, that his ashes be distributed.
>
> This package has been confined to your care for the
> fullfilment of this last will.
>
> You will kindly address a letter to Wm D. Haywood,
> Room 307, 164 W. Washington St., Chicago, Ill., telling
> the circumstances and where the ashes were distributed.
>
> **We Never Forget**
>
> JOE HILL MEMORIAL COMMITTEE.

Instructions distributed with envelopes containing Joe Hill's ashes. (*Courtesy of National Archives and Records Administration*)

ciety alerted the United Auto Workers Union, which investigated the find and published an article about Joe Hill's rediscovered remains in the April 1988 issue of its magazine *Solidarity*. From here the news soon spread to what was left of the International Workers of the World.

Like Joe Hill, very little remained of the IWW by 1988. With three hundred members worldwide, a two-room headquarters in Chicago, and one part-time employee to edit its paper, *The Industrial Worker* (circulation three thousand), it was the mere ashes of its former self. Big Bill Haywood was long gone, having died on a trip to Moscow in 1928; half of his ashes were deposited with fanfare in the Kremlin wall, and the rest were returned to Chicago and buried near the graves of his comrades who were executed for the Haymarket riot of 1886.

On June 1, 1988, Dr. Frederick S. Lee, chairman of the IWW General Executive Board, wrote to the archivist of the United States formally claiming ownership of Joe Hill's ashes and requesting their return: "What I and the organization I represent would like is that the ashes be returned to us so that we can carry out the last wishes of Joe Hill."

This sounds like a reasonable and simple request only if you are not a government bureaucrat. The archivist cannot "de-accession" an item in his possession and give it to anyone who claims it, without legal justification and proof of ownership. After investigating the matter, however, the staff of the archives found an ingenious justification: human remains are not archival in nature, any more than an old shoe or a rubber band, and therefore cannot legally be kept in the archives! A formal declaration was drawn up solemnly declaring that Joe Hill's ashes were not a record; the envelope, however, is a record, and it remains in official custody.

As to ownership, little could be done to determine who legally owned the ashes of a man who died seventy-three years before; and no attempt was made to trace his relatives. Charles Gepford, the original addressee of the envelope of ashes, could not be identified, and, since the ashes had once been in the possession of the IWW and no one else claimed them, after four months of deliberation the National Archives and Records Administration agreed to transfer them to Dr. Lee.

A minimalist ceremony took place on November 18, 1988, only one day short of the anniversary of Joe Hill's execution on November 19. Dr. Lee was accompanied to the headquarters of the National Archives by Bruce "Utah" Phillips, a traveling folksinger, storyteller, former archivist of Utah, and active member of the Wobblies. After signing a simple transfer form, they left with a small jar containing the long-lost ashes.

The IWW still retains some of Joe Hill's ashes at its new headquarters in San Francisco. From time to time, very small pinches are given out for use in labor movement ceremonies, including one, ironically, in Utah, where poor Joe had stated emphatically that he did not want to be found dead.

Sorry, Joe.

CHAPTER 3 7

The Delayed Burial of Eleanor Marx

K arl Marx, the crotchety and bewhiskered father of modern communism, died and is buried in England, which in his day was the bastion of capitalism. His place of burial surprises most people, who generally assume that he is mummified somewhere in Russia like Lenin, with whom his name and face were inextricably linked by Soviet propaganda. But the two never met, Lenin being only thirteen when Marx died on March 14, 1883. Not only did Marx spend his last thirty-four years in England, writing his masterwork *Das Kapital* in the British Museum Library reading room, but he was buried in London's distinctly middle-class Highgate Cemetery, rather than among the working classes whose cause he espoused.

One of his daughters, Eleanor Marx Aveling, became famous in her own right. She was a political writer, orator, socialist, and feminist; she translated Ibsen and Flaubert, and was a friend of George Bernard Shaw. She died an early death, however, at the age of forty in 1898. Her funeral was well attended, and occasioned much more interest than that of her father fifteen years earlier. Her body was cremated, but for some reason her husband never claimed the cinerary urn containing her ashes. A friend took charge of it eventually and brought it to the headquarters of the Socialist party. There it remained, in a glass-fronted cupboard, for twenty-three years.

By 1920, the Socialists had become the Communist Party

Common grave of Karl Marx and his family, including Eleanor Marx Aveling, in Highgate Cemetery, London. (*Photo by Edwin Murphy*)

of Great Britain, and the following year they moved their offices to 16 King Street, Covent Garden, taking with them the ashes of Eleanor Marx. By then, Soviet-style communism was viewed as a serious threat to social stability in England, and the British communists were under increasing pressure from the government. They were planning to send Eleanor's ashes to Moscow for safekeeping when the police raided their headquarters on May 7, 1921. The police officers, yielding to the urgent pleas of the party members, did not seize Eleanor's urn, as is sometimes erroneously reported. Thus, Eleanor Marx's ashes were spared the fate that befell those of Joe Hill in the United States, and did not end up in the state archives. However, the raid seems to have disrupted the plan of sending them to Moscow, for they stayed in the possession of the British Communist party until 1956. After 1933, they were on display in the Lenin Room of the Marx Memorial Library on Clerkenwell Green, except for the war years, when they were hidden in a basement to protect them from German bombs.

In 1956, Eleanor's remains were buried at last. The shoddy grave of her father had long since become a popular place of pilgrimage for visiting communists, socialists, and cemetery tourists, but its overgrown, poorly marked, and out-of-the-way location presented difficulties to those who sought it in Highgate. So Karl, his wife, and a couple of other family members were moved to a more prominent and accessible spot in the cemetery, and a huge, squat funerary bust of the founder of communism was raised over their common grave. To avoid gawking bystanders or possible demonstrators, the translation was carried out at night. In addition, the urn containing Eleanor's ashes, having escaped abandonment by her husband, seizure by Scotland Yard, exile to Russia, and Nazi air raids, and after fifty-eight years on display, was taken to Highgate by party members and reverently inserted into the grave of her father and her family. There it remains, although few visitors are aware of it.

Losing D. H. Lawrence

The controversial and consumptive British writer D. H. Lawrence, author of *Lady Chatterley's Lover*, *Sons and Lovers*, and other works that scandalized his contemporaries, died on March 2, 1930, near Vence in southern France. His devoted German wife, Frieda, buried him simply in the small local cemetery. Except for a headstone decorated with a mosaic of the phoenix, Lawrence's personal symbol, Frieda did nothing fancy for the obsequies or the gravesite, since she considered this a temporary interment. She intended to remove his body to their ranch in Taos, New Mexico, as soon as she could.

More easily said than done! It took the widow five years to get her husband back to Taos. In the process, his remains had to run a bureaucratic gauntlet of red tape and hostility, were lost three times, and in the end were mixed in a ton of concrete to keep them from being stolen. The awful odyssey of D. H.'s ashes is an edifying, if extreme, instance of the mischances that can befall a poor chap *after the funeral.*

Before his funeral, however, during eighteen years of marriage to Frieda, D.H. Lawrence had traveled widely. In 1922, he was lured to Taos, New Mexico, by the aspiring culture maven Mabel Dodge Luhan. At that time, Taos was remote and backward, but the wealthy dilettante Mabel was determined to transform it into a spiritual, intellectual, and artistic mecca, with herself as reigning patroness. Her collection of celebrities included Georgia O'Keeffe, Dorothy Brett, and Willa Cather as well as Lawrence.

Surprisingly, both Lorenzo (D.H.'s nickname) and Frieda fell in love with Taos. During his final protracted and peripatetic sojourn in Europe, when it became increasingly evident that his tuberculosis was terminal, D.H. had longed to return to his beloved Kiowa Ranch in Taos. Thus, when he died, Frieda felt it her duty to translate his remains to their own property where, according to her, he wished to be buried. "Now I have one desire," she wrote in a letter, "to take him to the ranch and make a lovely place for him there. He wanted so much to go." Mabel Dodge Luhan was happy to get him back within her cultural orbit, alive or dead, but she claimed that Lorenzo had told her that he wanted his ashes scattered over the ranch, not buried beneath it. This disagreement eventually led to a ludicrous quarrel, but the question was moot as long as the dead author was still moldering in his grave at Vence. Getting him home was the first problem, and it was a big problem indeed.

Frieda spent almost three years litigating in England over Lorenzo's not-too-sizable estate. It would cost a lot of money to exhume D.H. and transport his remains to New Mexico, but Frieda was determined to pay for it herself rather than accept the proffered financial aid of her friends. Thus, when she finally returned to Taos in 1933, it was without Lorenzo's body. Instead, she brought Angelo Ravagli, her next husband.

Frieda had been smitten with Angelo long before D.H. died. When first they met in the 1920s, Angelo owned a house the Lawrences were renting in Italy. He was an army officer, dashing in his Italian lieutenant's uniform. His obliging and cheerful personality endeared him to everyone. Lorenzo, impressed by his practical competence, remarked prophetically, "This man would be useful on the ranch." After vacating his villa, Frieda always kept track of Lieutenant Ravagli; she renewed their acquaintance after her husband's death and convinced him to accompany her to Taos in 1933 when the business over Lorenzo's estate was finally settled.

By the autumn of 1934, at Frieda's request, the complaisant Angelo had begun constructing a little shrine for his predecessor's remains on a sunny hillside above the ranch. She next prevailed upon him to fetch Lorenzo's body from Europe. Off he went, by rail and ship, for what turned out to be a six-month ordeal.

First there were bureaucratic difficulties with the French authorities in Vence. Getting all the proper authorizations to exhume the body was wearisome. Finally, however, the body was dug up and transported to Marseille. Since the difficulties of shipping a corpse were too daunting even for the resourceful and industrious Ravagli, the body was reduced to ashes on March 13, 1935, in the crematorium of St. Pierre Cemetery.

The next challenge was getting permission to export the remains from France and import them to the United States. Angelo booked passage on the *Conte de Savoia*, scheduled to sail from Villefranche on April 4, but as the sailing date approached he was still getting the runaround from the American consul in Genoa. He enlisted the aid of Frieda's daughter, Barbara, who tried to get around the problem by an appeal to the United States consul in London. When he finally went aboard the steamer with the urn, Ravagli was still not sure he had proper permission to land his little cargo in New York.

Apparently he hadn't. When he arrived in Manhattan on April 11, U.S. Customs and Immigration officials refused entry to the ashes of D. H. Lawrence. The government, which considered the English writer obnoxious to public morals (his works were banned), seemed intent on continuing its feud with him even beyond the grave. Protracted negotiations, and the intervention of Alfred Steiglitz, the photographer and art impresario, were necessary before the dangerous urn was grudgingly released from official custody.

That's when Angelo lost D.H. for the first time. The urn was misplaced for a while in a New York customs shed, but with Steiglitz's help it was finally located and turned over to the indefatigable and loyal Italian. Now all that was left was the gruelling four-day train trip to New Mexico. However, the real comedy was just beginning.

Angelo and the urn arrived hours late at Lamy, the nearest train station to Taos. Frieda and some friends came by car to meet him. Amid much animated greetings and excited conversation, the luggage was quickly piled in the auto for the long drive back to Taos. Unfortunately, one small item was left behind on the platform—the urn. Nobody missed it until they had covered twenty miles of dusty road. Lorenzo had been lost again! Back they went, and found the ashes right where they had been abandoned hours before.

By now it was too late to get to Taos before dark, so, after a festive dinner party at La Fonda Hotel in Santa Fe, they decided to pass the rest of the night paying a visit to a friend who lived en route. According to Dorothy Brett, Lawrence's friend and eccentric artist-in-residence at Taos, the house belonged to Nicolai Fechin, a Russian artist who was another of the celebrities Mabel Dodge Luhan had enticed to Taos.

Unfortunately, everyone was a little tired after the party, and when they finally arrived at Kiowa Ranch the next morning they realized that Lorenzo's urn was missing again! This was the third time in a week that D.H. had been lost. Most sources imply that the ashes were recovered the next day, but in a letter to Steiglitz, dated July 19, 1935, Dorothy Brett confides that they were left "for a whole week at Mrs. Fechin's house—she being Russian—emotional—made a little altar for them under an Icon and put candles round them and she says a great peace came from them, a great radiance, and she says from those ashes for the first time she really knew Lawrence."

There seems to have been some delicacy involved in Frieda's regaining possession of her thrice-misplaced husband, since the Fechins thought he had been left with them as a gift.

In any case, Lawrence finally made it back to Kiowa Ranch—or did he? Angelo Ravagli supposedly confessed years later to a visitor, the baron Prosper de Halleville, that to avoid problems and imagined customs fees for transporting human remains, he had jettisoned D.H.'s ashes before leaving France and refilled the empty urn with ordinary ashes after reaching New York. Although there are a lot of problems with this supposed confession as it stands, especially considering the difficulties Angelo encountered with officialdom all along the way, it is possible that at some point the harassed lieutenant actually may have cut some Gordian knot of bureaucratic red tape, in France or America, by quietly dumping the contents of the urn and replacing them later with ersatz Lorenzo dust. *Something* mysterious, which has never come to light, certainly happened to the ashes in New York. In a letter about it to Dorothy Brett on June 28, 1935, Alfred Steiglitz hinted, "Some day I'll tell you the whole story. Nothing quite like it has ever happened. Angelo really has no idea of what did happen."

Other unconfirmed legends have grown up in Taos that whatever ashes Angelo finally brought to New Mexico were accidentally spilled into the food by the writer Witter Bynner during the late-night welcome-home party at La Fonda Hotel. He furtively refilled the jar with fireplace ashes. These false remains (it is claimed) were secretly scattered to the winds by Mabel Luhan and replaced by ashes from her hearth. Later, Dorothy Brett filched these doubly fake ashes, scattered them at the ranch, and surreptitiously substituted powdery piñon dust from her stove. Brett, until her death at over ninety years old, would never decisively confirm or deny this story.

Unsubstantiated stories such as these easily sprouted as embellishments to the actual denouement concerning the ashes, which was farcical enough. There are several confusing and somewhat conflicting variations on the story, depending on whose letters, memoirs, autobiographies, and interviews you choose to believe, how you choose to combine them, and in which order, but sifting the evidence gives a fairly coherent picture along these lines:

Old rivalries among Frieda, Dorothy Brett, and Mabel Dodge Luhan for D. H. Lawrence's heart, mind, and soul, which abated when he left Taos in 1925 and were suppressed when he died in 1930, were rekindled as soon as Frieda revealed her plan to bring home his remains and enshrine them on a hillside in Taos. To Frieda, it was doing him honor and fulfilling his wish to be buried at his ranch.

To Brett and Luhan, however, the plan was crass, unworthy of Lorenzo's spirit and genius, and counter to what they claimed was his desire, which was to have his ashes scattered to the winds. Brett wrote to a friend, "Angelino has built this horrid chapel, this ghastly Villa d'Este affair of grass steps and fir trees and huge poles with a flame on top, with a cattle guard. It is so vulgar everyone is disgusted and laughing. . . ." She asked another friend, "Is a Mausoleum looking like a station toilet a fitting resting place?" To Steiglitz she wrote, "Shall I steal the ashes and place them in a tree where they will never be found?" Mabel considered the chapel burial, with its visitors' book, little better than a tourist trap; she also resented not being in charge of, or even being consulted on, the staging of the obsequies. She sensed a threat to her cultural dominance of Taos.

She and Brett hereupon contemplated stealing the ashes from the uncompleted chapel, before the grand interment ceremony, and release them to the winds over Taos. Brett wrote later:

> Several of us were involved in the plot, which seemed more of a game than a plot, as I don't believe any of us took it very seriously. . . . The plan was for me to take the

box with the ashes, and then they would be scattered. Most of it, believe me, was talk, the main fun was plotting the ways and means.

Mabel made the mistake of approaching Frieda's daughter, Barbara, who was visiting for the occasion. Barby revealed the plot to her mother (who may already have gotten wind of it from others). Mabel, sensing that the conspiracy was unraveling, sent a tart note to Frieda disputing her right to the ashes and threatening to wait until Frieda died, if necessary, to claim the ashes and scatter them as Lorenzo had wished. Mabel wrote to her friend Una Jeffers in September 1935:

Barby [told Frieda] all about the ashes project! Only saying it was my idea. . . ! Horrors! I racked and racked my brains but I could not recall telling Barby but I must have . . . Frieda sent insulting letters to us all. She . . . had Brett for three hours and used third degree and police methods, trying to get her to admit the thing.

Frieda and Angelo immediately ousted Brett from the ranch, where, despite her ridicule of the "mausoleum," she was employed in decorating it. They first posted a guard on the urn, but soon found a more permanent safeguard. Late one night, they mixed the dust of D. H. Lawrence into about a ton of sand and cement to form a massive block, which was used as the "altar" in the memorial shrine. Despite alleged threats by the Luhan faction to blow up the concrete structure, D. H. Lawrence is still there, neither buried on his ranch nor scattered all over it; he has become his own funerary monument! It will be hard to lose him anymore.

CHAPTER 39

Dorothy Parker: Excuse My Dust

Many New York lawyers undoubtedly have skeletons in their closets, but the firm of O'Dwyer and Bernstein, at 99 Wall Street, actually had Dorothy Parker's ashes in one of their file cabinets for twenty-one years. They didn't particularly want them, but they didn't know how to get rid of them because Dorothy herself left no instructions, her friends and executors washed their hands of them, and nobody else would take them until 1988, when they were finally buried with dignity in the headquarters of the National Association for the Advancement of Colored People (NAACP).

Dorothy Parker, born in 1893, was well known as a writer, book reviewer, and literary critic in the 1920s and 1930s. She also earned the reputation of being the wittiest woman in America. She was the only woman member of the weekly gathering of New York literati and intellectuals known as the Algonquin Round Table, so named from their meeting place in the Algonquin Hotel. A liberal all her life, Mrs. Parker, who was Jewish, died on June 7, 1967 and bequeathed her small estate of about $20,000, and her literary property rights, to Martin Luther King, Jr. Although she had never met him, she was impressed by his fervor in the cause of civil rights.

The Reverend Dr. King was surprised and touched by Parker's spontaneous generosity, but he never had a chance to put her money to good use before he was assassinated in 1968. Dorothy's will stipulated that in case of his death, her money and literary property would pass to the NAACP, and this took place when her estate was finally settled. By 1988, the estate was still producing up to $10,000 per year in income. But Dorothy Parker, who was so careful to provide for every contingency that might affect her estate, neglected to make such meticulous arrangements for herself. She suggested her own epitaph, "Excuse my dust," and asked for her body to be cremated, but she forgot to leave any further instructions.

She did insist that there be no funeral services of any kind, but her friend and executor, Lillian Hellman, ignored her wish and arranged a simple memorial ceremony, in which she and the comic actor Zero Mostel delivered eulogies. Lillian then sent her friend's body to be cremated the next day at Ferncliff Crematory in Hartsdale, New York. However, neither she nor anyone else showed up to claim Parker's ashes, and after a week the establishment threatened to dispose of them if no storage fees were forthcoming. No such fees being authorized from the estate, someone finally notified the crematorium to mail the orphaned ashes to the law firm of O'Dwyer and Bernstein on Wall Street, which was handling the estate settlement for Hellman. And there they stayed. Lillian Hellman, who lived for seventeen more years, never got around to telling the lawyers what to do with Mrs. Parker's remains.

After Hellman died, Paul O'Dwyer, the surviving law partner of the firm, began to worry about how to suitably provide for the final repose of Dorothy Parker. A meeting of prominent writers and others at the Algonquin Hotel in March 1988 considered several suggestions. Both the Algonquin Hotel and the *New Yorker* magazine, to which Parker had long contributed, declined to give the ashes a home. Other ideas ranged from scattering them over the Hudson River to mixing them in paint and then painting something, Dorothy's portrait for instance. But Benjamin Hooks, executive director of the NAACP, had the best solution. He offered to inter Dorothy Parker's ashes in a new memorial garden dedicated to her at his organization's headquarters in Baltimore, in token of thanks for her generosity and as a demonstration of political solidarity between blacks and Jews.

This congenial proposal was put into effect on October 20, 1988, with O'Dwyer, Hooks, the mayor of Baltimore, and many black and Jewish dignitaries in attendance. Dorothy Parker's ashes, in a gold-colored urn, were interred among newly planted trees under a marker inscribed: DEFENDER OF HUMAN AND CIVIL RIGHTS. The inscription also, however, included the words Dorothy herself had suggested: *Excuse My Dust*.

BIBLIOGRAPHY

NOTE: The bibliography has two parts. Sources relevant to only one person are listed first, in alphabetical order of the persons concerned. Second, general sources containing material on two or more people are listed under the heading "Miscellaneous" and alphabetized in the usual way.

BARRYMORE, JOHN

FLYNN, ERROL. *My Wicked, Wicked Ways.* G. P. Putnam's Sons. New York, 1959.

HAY, PETER. *Movie Anecdotes.* Oxford University Press. New York, 1990.

HENREID, PAUL. *Ladies Man: An Autobiography.* St. Martin's Press. New York, 1984.

KOBLER, JOHN. *Damned in Paradise: The Life of John Barrymore.* Atheneum. New York, 1977.

WALSH, RAOUL. *Each Man in His Time.* Farrar, Straus and Giroux. New York, 1974.

BOONE, DANIEL

FARAGHER, JOHN MACK. *Daniel Boone: The Life and Legend of an American Pioneer.* Henry Holt and Co. New York, 1992.

BUCKLAND, FRANK

"Sketch of Frank Buckland." *Popular Science Monthly.* January 1886, pp. 400–6.

BYRON, LORD (GEORGE GORDON)

LONGFORD, ELIZABETH. *The Life of Byron.* Little, Brown and Co. Boston, n.d.

MAUROIS, ANDRÉ. *Byron.* D. Appleton and Co. New York, 1930.

MAYNE, ETHEL COBURN. *The Life and Letters of Anne Isabella, Lady Noel Byron*. Dawsons of Pall Mall. London, 1969.

NICOLSON, HAROLD. *Byron: The Last Journey, April 1823–April 1824*. Houghton Mifflin Co. Boston, 1924 (reprint: Scholarly Press. St. Clair Shores, MI, 1972).

CASTRO, INEZ DE

BABELON, JEAN. "Inès de Castro." *Les femmes célèbres*. Editions d'art. Lucien Mazenod. Paris, 1960.

BLACK, HARMAN. *A Queen After Death*. The Real Book Co. New York, 1933.

CORNIL, SUZANNE. "Inès de Castro: Contribution à l'étude du développement littéraire du thème dans les littératures romanes; De l'histoire à la légende et de la légende à la littérature." *Mémoires*. Académie Royale de Belgique, Classe des lettres et des sciences morales et politiques. Tome XLVII, Fascicule 2, No. 1633, 1952.

Grande Enciclopédia Portuguesa e Brasileira. Editorial Enciclopédia, Limitada. Lisboa e Rio de Janeiro (n.d.); articles: "Castro, Inez de," vol. 6, and "Pedro I," vol. 20.

Lexicon des Mittelalters. Vol. 2, article: "Castro, Inês Pires de," pp. 1568–69. Artemis Verlag. Munich, 1983.

LIVERMORE, H. V. *A History of Portugal*. The University Press. Cambridge, MA, 1947.

LOPES, FERNÃO. *Chronica de El-Rei D. Pedro I*. Escriptorio. Lisbon, 1895.

MICHAUD, J. FR. *Biographie Universelle*. Vol. 20, pp. 331–32; article: "Inez de Castro." Paris, 1854.

PILLEMENT, GEORGES. *Unknown Portugal*. English translation by Arnold Rosin. Johnson Publications, Ltd. London, 1967.

SOCIÉTÉ DE SAVANTS ET DE GENS DE LETTRES. *La Grande Encyclopédie*. Vol. 9, p. 779; "Castro, Inez de." Paris, n.d (1898?).

WHEELER, DOUGLAS L. *Historical Dictionary of Portugal*. The Scarecrow Press. Metuchen, NJ, 1993.

CODY, "BUFFALO" BILL

DELLINGER, JOHN. "Buffalo Bill's Place in Old West History Is Rock Solid, Just Like His Colorado Mountain Burial Site." *Wild West*. October 1994, pp. 74–82.

WALSH, RICHARD J. *The Making of Buffalo Bill*. The Bobbs-Merrill Co. New York, 1928.

YOST, NELLIE SNYDER. *Buffalo Bill: His Family, Friends, Fame, Failure and Fortunes.* The Swallow Press. Chicago, 1979.

COLUMBUS, CHRISTOPHER

ARNESTO, FELIPE FERNANDEZ. *Columbus and the Conquest of the Impossible.* Saturday Review Press. New York, 1974.

BARMANN, GEORGE J. "Where Is Columbus Now?" *The American Mercury.* October 1957, pp. 92–94.

COLLINGWOOD, HARRIS, ed. "Maybe He Should Have Ordered a Bud Light." *Business Week.* March 13, 1989, p. 46.

DEHAINAUT, RAYMOND K. "Columbus Memorial: Another Invasion." *The Christian Century.* July 29–August 5, 1992, pp. 704–5.

FRENCH, HOWARD W. "For Columbus Lighthouse, a Fete That Fizzled." *New York Times International.* September 25, 1992.

GRANZOTO, GIANNI. *Christopher Columbus.* Translated by Stephen Sartarelli. Doubleday & Co., Inc. Garden City, NY, 1985.

LANGILLE, JAMES HILBERT, AND MARY F. FOSTER. *Popular History of the Life of Columbus.* Publications Bureau: Woman's National Press Association. Washington, DC, 1893.

MATHEWS, JAY. "IN QUEST OF THE TRUE COLUMBUS." *Washington Post,* October 13, 1986, pp. A-1, F-1.

SPENGLER, EUSEBIO LEAL, Official Historian of the City of Havana. Letter to author, February 20, 1990, with undated, untitled research paper on Columbus' burial in Havana (in Spanish).

TARDUCCI, FRANCESCO. *The Life of Christopher Columbus.* Translated by Henry F. Brownson. 2 vols. H. F. Brownson. Detroit, 1890.

TAVIANI, PAOLO EMILIO. *La Meravigliosa Adventuri di Cristoforo Columbo.* Instituto Geografico de Agostini, 1989.

"Where Lies Columbus?" *Time.* January 13, 1961, p. 30.

WILFORD, JOHN NOBLE. *The Mysterious History of Columbus: An Exploration of the Man, the Myth, the Legacy.* Alfred A. Knopf. New York, 1991.

WINSOR, JUSTIN. *Christopher Columbus.* Houghton Mifflin & Co. Cambridge, MA, 1892.

YOUNG, FILSON. *Christopher Columbus and the New World of His Discovery.* 2 vols. J. B. Lippincott Co. New York, 1906.

CROMWELL, OLIVER

FRASER, ANTONIA. *Cromwell: The Lord Protector.* Weidenfeld & Nicolson. London, 1973.

MOULD, RICHARD F. *Mould's Medical Anecdotes.* Adam Hilger, Ltd. Bristol, England, 1984.

EINSTEIN, ALBERT

"Einstein's Brain." *Omni.* December 1978, p. 40.

HARRISON, ERIC. "A Small-Town Doctor's Minding Einstein's Brain." *Los Angeles Times.* January 8, 1990, p. A4.

LEVY, STEVEN. "My Search for Einstein's Brain." *New Jersey Monthly.* August 1978, pp. 43–49.

MCCARTNEY, SCOTT. "The Hidden Secrets of Einstein's Brain Are Still a Mystery." *Wall Street Journal.* May 5, 1994, p. A-1.

MARANTO, GINA. "Einstein's Brain." *Discover,* May 1985, pp. 29–34.

MARSTON, WENDY. "Brain Teaser. TV 'Science' Plays a Cynical Joke." *The Sciences.* July/August 1994, p. 48.

METZGER, CLAIRE. "Einstein's Brain." *Fate.* September 1994, p. 57.

"Obsessions: Einstein's Brain." *The Economist.* April 2, 1994, p. 82.

REICH, WALTER. "The Stuff of Genius." *New York Times Magazine.* July 28, 1985, pp. 24–25.

"The Secret of Einstein's Area 39." *Omni.* August 1985.

WADE, NICHOLAS. "Einstein's Papers, and Brain: The Physicist's Legacy Remains Veiled to Millions." *New York Times.* July 27, 1987, p. A18.

GAGE, PHINEAS

BLAKESLEE, SANDRA. "Old Accident Points to Brain's Moral Center." *New York Times.* May 24, 1994, pp. C1, C14.

DAMASIO, HANNA, et al. "The Return of Phineas Gage: Clues About the Brain from the Skull of a Famous Patient." *Science.* May 20, 1994, pp. 1102–5.

"Neurology: Diagnosing A 146-Year-Old Injury." *Washington Post.* May 23, 1994, p. A2.

HARDY, THOMAS

HARDY, FLORENCE EMILY. *The Later Years of Thomas Hardy, 1892–1928.* The MacMillan Co. New York, 1930.

MILLGATE, MICHAEL. *Thomas Hardy: A Biography.* Random House. New York, 1982.

PINION, F. B. *Thomas Hardy: His Life and Friends.* St. Martin's Press. New York, 1992.

HAYDN, JOSEPH

BUTTERWORTH, NEIL. *Haydn: His Life and Times.* Midas Books. Tunbridge Wells, Kent, 1977.

"Haydn's Skull Is Returned: After Theft 145 Years Ago, Body Is Complete." *Life.* June 28, 1954, pp. 51, 52, 54.

LANDON, H. C. ROBBINS, AND HENRY RAYNOR. *Haydn.* Praeger Publishers. New York, 1972.

"Skull Is Restored to Haydn's Grave." *New York Times.* June 6, 1954, p. 6.

STATLAENDER, CHRISTINA. *Joseph Haydn of Eisenstadt.* Translated by Percy M. Young. Dennis Dobson. London, 1968.

WESCHBERG, JOSEPH. "Our Far-flung Correspondents: An E in the Seventh Bar." *The New Yorker.* May 15, 1954, pp. 100 et seq.

HILL, JOE

CARSON, PETER. *Roughneck.* W. W. Norton & Co. New York, 1983.

CHAPLIN, RALPH. *Wobbly: The Rough-and-Tumble Story of an American Radical.* University of Chicago Press. Chicago, 1948.

FOGEL, CHUCK. "Joe Hill's Ashes." *Solidarity* (published by United Automobile Workers Union). April 1988, p. 23.

KERNAN, MICHAEL. "Joe Hill Remembered: The Activist's Ashes Are Given to Union Leaders." *Washington Post.* November 19, 1988, pp. C1, C6.

National Archives correspondence file on Joe Hill, File 47388; Records Relating to the Espionage Act, World War I, 1917–21; Records of the Post Office Department, Office of the Solicitor; Record Group 28; National Archives, Washington, D.C.

SMITH, GIBBS M. *Joe Hill.* University of Utah Press. Salt Lake City, 1969.

South, Aloha, senior archivist. Interview with author. National Archives and Records Administration (Pennsylvania Avenue). Washington, D.C., June 1994.

STEGNER, WALLACE. *Joe Hill: A Biographical Novel.* University of Nebraska Press. Lincoln, Nebraska, 1980.

"WOBBLY." *The New Yorker.* Dec. 19, 1988, p. 28.

JONES, JOHN PAUL

BUELL, AUGUSTUS C. *Paul Jones: Founder of the American Navy.* Books for Libraries Press. Freeport, NY, 1900.

FOX, JOSEPH L. *Captain John Paul Jones: Forgotten Naval Hero.* Adams Press. Chicago, 1987.

JOHNSON, GERALD W. *The First Captain: The Story of John Paul Jones.* Coward-McCann, Inc. New York, 1947.

LORENZ, LINCOLN. *John Paul Jones: Fighter for Freedom and Glory.* United States Naval Institute. Annapolis, MD, 1943.

MORRISON, SAMUEL ELIOT. *John Paul Jones: A Sailor's Biography.* Little, Brown and Company. Boston, 1959.

RUSSEL, PHILIPS. *John Paul Jones: Man of Action.* Brentano's. New York, 1927.

JONSON, BEN

REES, NIGEL. *Epitaphs: A Dictionary of Grave Epigrams and Memorial Eloquence.* Caroll & Graf Publishers. New York, 1993.

LAWRENCE, D. H.

BRETT, DOROTHY. *Lawrence and Brett: A Friendship.* New edition, with introduction, prologue and epilogue by John Manchester. The Sunstone Press. Santa Fe, New Mexico, 1974.

BYNNER, WITTER. *Journey With Genius: Recollections and Reflections Concerning the D. H. Lawrences.* The John Day Co. New York, 1951.

HAHN, EMILY. *Mabel: A Biography of Mabel Dodge Luhan.* Houghton Mifflin Company. Boston, 1977.

HIGNETT, SEAN. *Brett: From Bloomsbury to New Mexico.* Franklin Watts. New York, 1983.

LUCAS, ROBERT. *Frieda Lawrence: The Story of Frieda Von Richthofen and D. H. Lawrence.* Translated by Geoffrey Skelton. Secker & Warburg. London, 1973.

MOORE, HARRY T. *The Priest of Love: A Life of D. H. Lawrence.* (Previously published as *The Intelligent Heart*). Farrar, Straus and Giroux. New York, 1974.

MOORE, HARRY T., AND DALE B. MONTAGUE, ed. *Frieda Lawrence and Her Circle: Letters from, to and about Frieda Lawrence.* Archon Books. Hamden, CT, 1981.

MOORE, HARRY T., AND WARREN ROBERTS. *D. H. Lawrence.* Thames & Hudson. New York, 1988.

NEHLS, EDWARD, ed. *D. H. Lawrence: A Composite Biography.* University of Wisconsin Press. Madison, WI 1958.

RUDNICK, LOIS PALKEN. *Mabel Dodge Luhan: New Woman, New Worlds.* University of New Mexico Press. Albuquerque, NM, 1984.

"Taos." *Washington Post,* October 16, 1988, pp. E1, E8.

TEDLOCK, E. W., JR., ed. *Frieda Lawrence: The Memoirs and Correspondence.* Alfred A. Knopf. New York, 1964.

LINCOLN, ABRAHAM

COHEN, GARY. "Back to the Future." *U.S. News & World Report.* April 25, 1994, p. 27.

KUBICEK, EARL C. "The Lincoln Corpus Caper." *Lincoln Herald.* Fall 1980, pp. 474–80.

KUNHARDT, DOROTHY MESERVE. "Saga of Lincoln's Body." *Life.* February 15, 1963, pp. 83–88.

MCMURTRY, R. GERALD. "The Attempt to Steal Lincoln's Body." *Lincoln Lore.* March 1972, pp. 1–2.

———. "Viewing Lincoln's Remains." *Lincoln Lore.* May 1972, pp. 2–3.

MERIAM, ARTHUR L. "Final Interment of President Abraham Lincoln's Remains at the Lincoln Monument in Oak Ridge Cemetery, Springfield, Illinois." *Journal of Illinois State Historical Society.* April 1930, pp. 171–4.

ROBERTSON, DEANE, AND PEGGY ROBERTSON. "The Plot to Steal Lincoln's Body." *American Heritage.* 33:3 (1982), pp. 76–83.

TULEJA, TAD. *Curious Customs: The Stories Behind 296 Popular American Rituals.* Harmony Books. New York, 1987.

WARREN, LOUIS A. "The Plot to Steal Lincoln's Corpse." *Lincoln Lore.* June 12, 1944 (reprint, May 1972, pp. 1–2).

LIVINGSTONE, DAVID

EATON, JEANETTE. *David Livingstone: Foe of Darkness.* William Morrow & Co. New York, 1947.

HELLY, DOROTHY O. *Livingstone's Legacy: Horace Walter and Victorian Mythmaking.* Ohio University Press. Athens, OH, 1987.

HUXLEY, ELSPETH. *Livingstone and His African Journeys.* Saturday Review Press. New York, 1974.

JEAL, TIM. *Livingstone.* G. P. Putnam's Sons. New York, 1973.

MACNAIR, JAMES I. *Livingstone the Liberator: A Study of a Dynamic Personality.* Collins Clear-Type Press. London, 1953.

SEAVER, GEORGE. *David Livingstone: His Life and Letters*. Harper & Brothers, Publishers. New York, 1957.

MARX, ELEANOR

KAPP, YVONNE. *Eleanor Marx*. Pantheon Books. New York, 1976.

MOLIÈRE

CHATFIELD-TAYLOR, H. C. *Molière*. Duffield & Co. New York, 1906.

CHAUMEROT. *A Fortnight in Paris: Chaumerot's New Illustrated Pocket Guide*. Paris, 1852.

A. AND W. GALIGNANI AND CO. *Galignani's New Paris Guide*. Paris, 1842.

L'Intermediaire des chercheurs et curieux. LXII (1900), columns 584, 681–3, 744.

JACOB, P. L. *"Quelques notes sur le tombeau et sur le cercueil de Molière."* *Moliériste*. VI (1884–5), pp. 131–6.

MANDER, GERTRUD. *Molière*. English translation by Diana Stone Peters. Frederick Ungar Publishing Co. New York, 1973.

MARZIALIS, FRANK T. *Molière*. George Bell & Sons. London, 1906.

MOLAND, LOUIS. *"La sépulture ecclésiastique donnée à Molière."* *Moliériste*. VI (1884–5), pp. 67–81.

MONVAL, GEORGES. *"Les tombeaux de Molière et de la Fontaine."* *Moliériste*. IV (1882–3), pp. 3–9.

PALMER, JOHN LESLIE. *Molière*. Benjamin Blom, Inc. New York, 1970.

VAN LOON, HENDRIK WILLIAM. *The Arts*. Simon & Schuster. New York, 1937.

MOZART, WOLFGANG AMADEUS

BAHN, PAUL G. *"The Face of Mozart."* *Archaeology*. March/April 1991, pp. 38–41.

BRAUNBEHRENS, VOLKMAR. *Mozart in Vienna, 1781–1791*. English translation by Timothy Bell. Harper Perennial. New York, 1991.

GLAUSIUSZ, JOSIE. *"The Banal Death of a Genius."* *Discover*. March 1994, p. 25.

NOBBE, GEORGE. *"The Fall of Mozart."* *Omni*. December 1994, p. 36.

WISE, MICHAEL Z. *"Anatomy of a Brain Teaser: In Austria, the Contested Skull of Mozart."* *Washington Post*. November 22, 1991, p. B2.

NAPOLEON

AUBREY, OCTAVE. *St. Helena.* Translated by Arthur Livingston. J. B. Lippincott Co. London, 1936.

JONES, PROCTOR PATTERSON. *Napoleon: An Intimate Account of the Years of Supremacy, 1800–1814.* Random House. New York, 1992.

"Napoléon: arsenic et vieilles polémiques." *Lyon Matin.* August 7, 1994, p. 69.

WAXMAN, SHARON. "A Brush with Napoleon: Jean Fichou Finds It Hard to Part With Lock of Napoleon's Hair." *Washington Post.* June 11, 1994, pp. B1–B2.

PAINE, THOMAS

AYER, A. J. *Thomas Paine.* Atheneum. New York, 1988.

BEST, MARY AGNES. *Thomas Paine: Prophet and Martyr of Democracy.* Harcourt, Brace & Company. New York, 1927.

BRESSLER, LEO A. "Peter Porcupine and the Bones of Tom Paine." *The Pennsylvania Magazine of History and Biography.* Vol. LXXXII, No. 2. April 1958, pp. 176–85.

DYCK, IAN, ed. *Essays on Thomas Paine.* St. Martin's Press. New York, 1988.

EDWARDS, SAMUEL. *Rebel! A Biography of Tom Paine.* Praeger Publishers. New York, 1974.

FONER, PHILIP S. *The Life and Major Writings of Thomas Paine.* Citadel Press. New York, 1993.

GURKO, LEO. *Tom Paine: Freedom's Apostle.* Thomas Y. Crowell Company. New York, 1957.

HAWKE, DAVID FREEMAN. *Paine.* W. W. Norton. New York, 1992.

POWELL, DAVID. *Tom Paine: The Greatest Exile.* St. Martin's Press. New York, 1985.

WILLIAMSON, AUDREY. *Thomas Paine: His Life, Work and Times.* George Allen & Unwin Ltd. London, 1973.

WOODWARD, W. E. *Tom Paine: America's Godfather.* E. P. Dutton & Co. New York, 1945.

PARKER, DOROTHY

"Dorothy's Dust." *U.S. News & World Report.* March 28, 1988, p. 11.

HALL, CARLA. "Dorothy Parker: Wit's End: Poet's Ashes, Ideals Honored in Baltimore." *Washington Post.* October 21, 1988, pp. D1–D2.

MEADE, MARION. *Dorothy Parker: A Biography.* Villard Books. New York, 1988.

FREWIN, LESLIE. *The Late Mrs. Dorothy Parker.* Macmillan Publishing Co. New York, 1968.

"Twenty-one Years Later: Remains of Dorothy Parker Go to NAACP Headquarters." *Jet.* April 11, 1988, p. 54.

PERÓN, EVA

BARNES, JOHN. *Evita, First Lady: A Biography of Eva Perón.* Grove Press. New York, 1978.

ESCOBAR, GABRIEL. " 'Evita': Cry, Argentina." *Washington Post.* June 23, 1994, pp. C1, C2.

FRASER, NICHOLAS, AND MARYSA NAVARRO. *Eva Perón.* W. W. Norton & Co. New York, 1980.

PERÓN, JUAN

"Case of the Severed Hands." *Time.* July 20, 1987, p. 51.

SHAPIRO, ARTHUR M. "Political Necrophilia: Holding (Perón's) Hands in Argentina." *The New Leader.* August 10–24, 1987, pp. 5–7.

"Thieves Ask $8 Million for Hands of Perón." *Washington Post.* July 3, 1987, p. A30.

RICHELIEU (ARMAND EMMANUEL DU PLESSIS, CARDINAL RICHELIEU)

BALLINI, A.-C. *Un bleu des Côtes-d'Armor: Nicolas Armez.* Spézet, 1990.

DUBREUIL, L. *Le fureteur breton.* Saint-Brieuc, 1909–1910: 4ème année, p. 189; 5ème année, pp. 22, 59 et suivantes.

———. *Un révolutionnaire de Busse-Bretagne: Nicolas Armez.* Paris, 1929.

KERLEVEO, MGR. *Paimpol et son terroir.* Rennes, 1971.

LABARRE DE RAILLICOURT, DOMINIQUE. *Richelieu: Le Maréchal Libertin.* Librairie Jules Tallandier. Paris, 1991.

O'CONNELL, DANIEL PATRICK. *Richelieu.* Weidenfeld and Nicolson. London, 1968.

PRICE, ELEANOR K. *Cardinal De Richelieu.* McBride, Nast & Co. New York, 1912.

STAVRIDÈS, YVES. *"Le Roman des Archives."* *L'Express International.* June 2, 1994, pp. 26–33.

WILKINSON, BURKE. *Cardinal in Armor*. The Macmillan Co. New York, 1966.

SHELLEY, PERCY BYSSHE

BUXTON, JOHN. *Byron and Shelley: The History of a Friendship*. Harcourt, Brace & World. New York, n.d.

CIANFARRA, CAMILLE. "Historic Cemetery in Rome Seeks Aid." *New York Times*. March 6, 1950, p. 5.

DOWDEN, EDWARD. *Life of Percy Bysshe Shelley*. Barnes & Noble reprint. New York, 1966 (originally published 1886).

GRYLLS, ROSALIE GLYNN. *Mary Shelley*. Oxford University Press. London, 1938.

HOLMES, RICHARD, ed. *Shelley: The Pursuit*. E. P. Dutton & Co. New York, 1975 .

PECK, WALTER EDWIN. *Shelley: His Life and Work*. Houghton Mifflin Co. Boston, 1927.

ROE, IVAN. *Shelley: The Last Phase*. Hutchinson. London, 1953.

TRELAWNY, EDWARD JOHN. *Recollections of Shelley and Byron*. Reprinted in *The Life of Percy Bysshe Shelley*, vol. 2. J. M. Dent & Sons. London, 1933.

WHITE, NEWMAN IVY. *Shelley*. Octagon Books. New York, 1972.

SITTING BULL

"Bones of Sitting Bull Go South From One Dakota to the Other." *New York Times*. April 9, 1953, p. 29.

BROWN, DEE. *Bury My Heart at Wounded Knee: An Indian History of the American West*. Holt, Rinehart & Winston. New York, 1971.

COLLINS, DABNEY OTIS. "The Fight for Sitting Bull's Bones." *American West 3*. Winter 1966, pp. 72–78.

"Dakotas in Clash Over Sitting Bull." *New York Times*. March 22, 1953, p. 50.

FONER, ERIC, AND JOHN A. GARRATY, ed. *The Reader's Companion to American History*. Houghton Mifflin Co. Boston, 1991.

HAMMER, KENNETH M. "Sitting Bull's Bones." English Westerns Society. *Brand Books 23*. Winter 1984, pp. 1–8.

"The Sitting Bull Argument." *New York Times*. April 10, 1953, p. 23.

"Sitting Bull in the News." *New York Times*. April 24, 1953, p. 22.

"Sitting Bull—Last Act?" *New York Times*. August 29, 1953, p. 19.

"Sitting Bull Rises Again." *New York Times*. December 19, 1953, p. 17.

"Sitting Bull's Bones Resist 'Raid,' North Dakota Says." *New York Times*. May 9, 1953, p. 60.

"Sorry, We Do Not Have Sitting Bull's Skull, Smithsonian Replies to Sioux Indian Plea." *New York Times*. February 13, 1953, p. 26.

UTLEY, ROBERT M. *The Lance and the Shield: The Life and Times of Sitting Bull*. Ballantine Books. New York, 1993.

THORPE, JIM

CARBON COUNTY TOURIST PROMOTION AGENCY. BROCHURE: *Jim Thorpe: The Man . . . And the Town*. Jim Thorpe, PA, 1993.

FORBES, CHARLOTTE. "Jim Thorpe Lines Up for Visitors." *Saturday Evening Post*. March 1990, pp. 86–87.

FUSSMAN, CAL. "An Interrupted Journey." *Washington Post Magazine*. March 14, 1993, pp. 22–5, 34–5.

GOBRECHT, WILBUR. J. *Jim Thorpe, Carlisle Indian*. Cleveland County Historical Society and Hamilton Library Association. Cleveland, OH, 1985.

METZGER, CLAIR. "Jim Thorpe's Tormented Spirit." *Fate*. June 1989, pp. 52–54.

NEWCOMBE, JACK. *The Best of the Athletic Boys*. Doubleday & Co. Garden City, NY, 1975.

OUTERBRIDGE, LAURA. "Town Thrives on an Athlete's Aura." *Insight*. June 20, 1988, pp. 24–25.

SCHOOR, GENE. *The Jim Thorpe Story: America's Greatest Athlete*. Julian Messner. New York, 1951.

VOLTAIRE

ANDREWS, WAYNE. *Voltaire*. New Directions Books. New York, 1981.

BRANDES, GEORGE. *Voltaire*. Tudor Publishing Co. New York, 1936.

HEARSEY, JOHN E. N. *Voltaire*. Barnes & Noble. New York, 1976.

NOYES, ALFRED. *Voltaire*. Sheed & Wood. New York, 1936.

ORIEUX, JEAN. *Voltaire*. English translation by Barbara Bray and Helen R. Lane. Doubleday & Co. Garden City, NY, 1979.

VULLIAMY, COLWYN EDWARD. *Voltaire*. Kennikat Press. Port Washington, NY, 1970.

WILLIAM THE CONQUEROR

DOUGLAS, DAVID C. *William the Conqueror; The Norman Impact Upon England*. The University of California Press. Berkeley, 1964.

LONGFORD, ELIZABETH, ed. *The Oxford Book of Royal Anecdotes*. Oxford University Press. Oxford, 1989.

VITALIS, ORDERICUS. *The Ecclesiastical History of England and Normandy*. Translated by Thomas Forester. 4 vols. Henry G. Bohn. London, 1853.

WILLIAMS, ROGER

CHAPIN, HOWARD M. "Odd Phases of Rhode Island History: Part of Apple Tree Root That Had Grown into Grave of Roger Williams and Dug Up in 1860 Now an Exhibit at Historical Society." *The Providence Journal* (Rhode Island). July 6, 1924, Fifth Section, p. 8.

EASTON, EMILY. *Roger Williams: Prophet and Pioneer*. AMS Press. New York, 1969.

ERNST, JAMES. *Roger Williams: New England Firebrand*. AMS Press. New York, 1969.

SCHATZ, ALBERT. "Recycling of Roger Williams." *The Providence Journal* (Rhode Island). April 12, 1973, p. 30.

MISCELLANEOUS

ANGELL, CHARLES BRADFORD. *Heart Burial*. George Allen & Unwin. London, 1933.

ARBEITER, JEAN, AND LINDA D. CIRINO. *Permanent Addresses: A Guide to the Resting Places of Famous Americans*. M. Evans and Co. New York, 1983.

BALL, JAMES MOORE. *The Sack-em-up Men: The Story of the Resurrectionists*. Oliver and Boyd. Edinburgh, 1928.

BARING-GOULD, SABINE. *Curious Myths of the Middle Ages*. New edition; Longmans, Green, and Co. London, 1897.

BARKER, FELIX. *Highgate Cemetery: Victorian Valhalla*. Salem House. Salem, NH, 1984.

BENCHLEY, NATHANIEL. *Humphrey Bogart*. Little, Brown & Co. Boston, 1975.

BENTLEY, JAMES. *Restless Bones: The Story of Relics*. Constable and Co. London, 1985.

BOAS, THOMAS S. R. *Death in the Middle Ages: Mortality, Judgment and Remembrance*. Thames and Hudson. London, 1972.

BOLLER, PAUL F., JR., AND RONALD L. DAVIS. *Hollywood Anecdotes*. William Morrow and Co. New York, 1987.

BORD, JANET, AND COLIN BORD. *Unexplained Mysteries of the 20th Century*. Contemporary Books. Chicago, 1989.

BOUDREAU, JOHN. "Couldn't You Just Die? Necropolis, USA: One Town's Underground Economy." *Washington Post*. June 12, 1994, pp. F1, F4.

BROWN, DAVID. "Uncovering a Therapy From the Grave." *Washington Post*. October 25, 1993, p. A3.

BUSHELL, PETER. *Great Eccentrics*. Unwin Paperbacks. London, 1985.

CROSS, DAVID, AND ROBERT BENT. *Dead Ends: An Irreverent Field Guide to the Graves of the Famous*. Plume. New York, 1991.

CULBERTSON, JUDI, AND TOM RANDALL. *Permanent Californians: An Illustrated Guide to the Cemeteries of California*. Chelsea Green Publishing Company. Chelsea, VT, 1989.

———. *Permanent New Yorkers: A Biographical Guide to the Cemeteries of New York*. Chelsea Green Publishing Company. Chelsea, VT, 1987.

———. *Permanent Parisians: An Illustrated Guide to the Cemeteries of Paris*. Chelsea Green Publishing Company. Chelsea, VT, 1986.

CURL, JAMES STEVENS. *The Victorian Celebration of Death*. The Partridge Press. Detroit, 1972.

DICKERSON, ROBERT B. *Final Placement: A Guide to the Deaths, Funerals, and Burials of Famous Americans*. Reference Publications. New York, 1982.

DONALDSON, NORMAN. *How Did They Die?* Vol. 3. St. Martin's Press. New York, 1994.

DONALDSON, NORMAN, AND BETTY DONALDSON. *How Did They Die?* Vols. 1 and 2. St. Martin's Press. New York, 1980.

DUPUY, R. ERNEST, AND TREVOR N. DUPUY. *The Encyclopedia of Military History: From 3500 B.C. to the Present*. Second revised edition. Harper & Row. New York, 1986.

EDWARDS, FRANK. *Strange People*. Citadel Press. Secaucus, NJ, 1961.

———. *Stranger Than Science*. Citadel Press. Secaucus, NJ, 1959.

———. *Strangest of All*. Citadel Press. Secaucus, NJ, 1956.

Encyclopedia Britannica. Chicago, 1957. Various articles.

ELLIS, NANCY, AND PARKER HAYDEN. *Here Lies America: A Collection of Notable Graves*. Hawthorn Books. New York, 1978.

237

EXTON, PETER AND DOROTHY KLEITZ. *Milestones into Headstones: Minibiographies of Fifty Fascinating Americans Buried in Washington, D.C.* EPM Publications. McLean, VA, 1985.

Facts and Fallacies. The Reader's Digest Association. Pleasantville, NY, 1988.

FORBES, MALCOLM, AND JEFF BLOCH. *They Went That-a-Way.* Simon & Schuster. New York, 1988.

FRAZER, JAMES GEORGE. *The Golden Bough.* Various editions.

GEARY, PATRICK J. *Furta Sacra: Thefts of Relics in the Central Middle Ages.* Revised edition. Princeton University Press. Princeton, NJ, 1990.

GOYA, FRED, AND MIKE MORIARTY. *What a Way to Go! A Compendium of Bizarre Demises.* Citadel Press. Secaucus, NJ, 1982.

HENDRICKSON, ROBERT. *The Literary Life and Other Curiosities.* Penguin Books. Harmondsworth, Middlesex, England, 1982.

HIGGINS, PAUL LAMBOURNE. *Pilgrimages: A Guide to the Holy Places of Europe for Today s Traveler.* Prentice-Hall. Englewood Cliffs, New Jersey, 1984.

JONES, BARRY, AND MEREDITH VIBART DIXON. *The St. Martin's Press Dictionary of Biography.* St. Martin's Press. New York, 1986.

LO BELLO, NINO. *Nino Lo Bello's Guide to Offbeat Europe.* Chicago Review Press. Chicago, 1985.

LONG, KIM, AND TERRY REIM. *Fatal Facts: A Lively Look at Common and Curious Ways People Have Died.* Arlington House. New York, 1985.

LOUIS, DAVID. *2201 Fascinating Facts.* Greenwich House. New York, 1983.

MALONE, MICHAEL. "The Grave Seekers: Literary Pilgrimage with a Twist: In Search of Poe's Tomb and Other Spots Where Writers Rest." *Washington Post.* October 25, 1987, pp. E1, E5.

MARION, JOHN FRANCIS. *Famous and Curious Cemeteries.* Crown Publishers. New York, 1977.

MORROW, ED. *The Grim Reaper's Book of Days: A Cautionary Record of Famous, Infamous, and Unconventional Exits.* Citadel Press. New York, 1992.

MOULD, RICHARD F. *More of Mould's Medical Anecdotes.* Adam Hilger. Bristol, England, 1989.

NASH, JAY ROBERT. *Zanies: The World's Greatest Eccentrics.* New Century Publishers. Piscataway, NJ, 1982.

The New Encyclopaedia Britannica. Fifteenth edition. Chicago, 1974. Various articles.

OLLOVE, MICHAEL. "Maryland Man Being Mummified: From a Baltimore Basement, Treatment Fit for a King." *The Baltimore Sun.* June 17, 1994, pp. 1A, 11A.

PANATI, CHARLES. *Panati's Extraordinary Endings of Practically Everything and Everybody.* Harper & Row. New York, 1989.

PLATNICK, KENNETH. *Great Mysteries of History.* Dorset Press. New York, 1987.

PUCKLE, BERTRAM S. *Funeral Customs: Their Origin and Development.* T. Werner Laurie, Ltd. London, 1926.

RANDALL, DAVID. *Royal Follies: A Chronicle of Royal Misbehavior.* Sterling Publishing Co. New York, 1988.

RATAZZI, PETER. *In Strangest Europe.* Mitre Press. London, 1968.

REICH, WALTER. "The Stuff of Genius." *New York Times Magazine.* July 28, 1985, pp. 24-25.

"Remains to Be Seen." *People Magazine.* December 28, 1987, p. 102.

RUFUS, ANNELI S., AND KRISTAN LAWSON. *Europe Off the Wall.* John Wiley & Sons. New York, 1988.

SOX, DAVID. *Relics and Shrines.* George Allen & Unwin. London, 1985.

Strange Stories, Amazing Facts. The Reader's Digest Association. Pleasantville, NY, 1976.

THORNE, J. O., ed. *Chambers's Biographical Dictionary.* New edition. St. Martin's Press. New York, 1962.

THURSTON, HERBERT, S. J. *The Physical Phenomena of Mysticism.* Burns, Oates. London, 1952.

WALKER, CHARLES. *Strange Britain.* Brian Todd Publishing House. London, 1989.

WALLACE, IRVING, et al. *Significa.* E. P. Dutton. New York, 1983.

WEIL, TOM. *The Cemetery Book: Graveyards, Catacombs, and Other Travel Haunts Around the World.* Barnes and Noble Books. New York, 1993.

WILSON, IAN. *Undiscovered.* Beech Tree Books/William Morrow. New York, 1987.

INDEX

Harvey, Thomas, 196–201
Havana, 92, 93, 94, 95
Haydn, Joseph, 7–12
Haywood, Big Bill, 205, 206, 209
Heads, 1–34
Hearts, 37–64, 97–106, 165, 169
Heel bones, 64, 101, 189
Hellman, Lillian, 222
Hell's Angels, 68–69
Henreid, Paul, 73
Henri III (France), 111
Henri IV (France), 100
Henry II (England), 58
Henry IV (England), 64
Highgate cemetery (London), 134, 211–13
Hill, Joe, 202–10
Hooks, Benjamin, 223
Houdini, Harry, 67
Hughes, Jack, 161, 162
Hull, Richard, 67–68
Hume of Godscroft, 61
Hundsthurm cemetery (Vienna), 9
Hunt, Leigh, 41, 42, 43, 45
Hyrtl, Jacob, 15
Hyrtl, Joseph, 15, 16

Industrial Workers of the World (IWW), 203–10
International Mozarteum Foundation (Salzburg), 13, 16–17, 18
Intestines, 58, 165, 189

Jeffers, Una, 220
Jim Thorpe, Pennsylvania, 181–86
John, King (England), 190
John XXII, Pope, 61
Jones, John Paul, 113–21
Jonson, Ben, 64

Keats, John, 39, 45
Keats-Shelley Memorial (Rome), 43
Kensal Green cemetery (London), 134
Keratry, Auguste de, 26
Kinelly, "Big Jim," 160–62
King, Martin Luther, Jr., 221
Kipling, Rudyard, 55
Kirk, John, 170
Krueger, Otto, 151

La Bosse, 103
Lafayette, Marquis de, 49
La Fontaine, Jean de, 108, 110
Larkin, James, 207
Laurel Hill cemetery (Philadelphia), 134
Lawrence, Barbara, 216, 220
Lawrence, D. H., 214–20
Lawrence, Frieda, 214–20
Lee, Frederick S., 209, 210
Lemaître, 101
Lennox Library (New York), 95
Lenoir, Alexandre, 26
Lenoir's museum, 110, 111
Leopold, Duke of Austria, 57
Levy, Steven, 196, 198, 200, 201
Library of Congress, 95
Lincoln, Abraham, 155–64
Lincoln, Mary, 159
Lincoln, Robert Todd, 161, 163
Lincoln, Willie, 158
Livers, 50
Livingstone, David, 165–71, 196
Lookout Mountain (Colorado), 143, 144–47
Lopez, Fernão, 84–86
Lorraine, Louise de, 111
Lorre, Peter, 73
Louis XIII (France), 23
Louis XIV (France), 64–64
Louis XVI (France), 102, 113, 116, 190
Luhan, Mabel Dodge, 214–20

MacDonald, Ramsay, 55
McLaughlin, James, 149
Malloy Brothers' mortuary (Los Angeles), 72
Malloy, Dick, 72
Marshall, John, 131
Marx, Eleanor, 211–13
Marx, Karl, 211–13
Marx Memorial Library (London), 213
Mauch Chunk, Pennsylvania, 181, 185
Mayer, Louis, 74
Mignot, Abbé, 98, 99
Miller, Charles, 71
Mithouart, Monsieur, 99, 102
Mobridge, South Dakota, 151–54
Molière, 107–12
Montalembert, Comte de, 27